HENTOPIA

Hentopia

Create a Hassle-Free Habitat *for* Happy Chickens

21 Innovative Projects

FRANK HYMAN

Storey Publishing

The mission of Storey Publishing is to serve our customers by publishing practical information that encourages personal independence in harmony with the environment.

Edited by Deborah Burns
Art direction and book design by Michaela Jebb
Text production by Erin Dawson
Indexed by Christine R. Lindemer, Boston Road Communications

Interior photography by © Liz Nemeth

Cover images by © AzFree/iStockphoto.com, spine (background); © Image Source/Getty Images, front & spine; © Katie Eberts, front (background); © Liz Nemeth, back

Additional photography by Courtesy of Frank Hyman, 8 bottom, 47 bottom, 71, 84, 87, 92, 140, 172, 188, 191, 193; © Grigorenko/iStockphoto.com, 115; © Image Source/Getty Images, 2; © kali9/iStockphoto.com, 10

Illustrations by © Michael Gellatly

This book is intended to help readers create devices and small structures for their backyard chickens. Be sure to read all explanations and advice in this book thoroughly before undertaking your own projects. Take proper safety precautions before using potentially dangerous tools and equipment or undertaking potentially dangerous activities. Be alert and vigilant while operating power equipment and using sharp tools.

Storey Publishing
210 MASS MoCA Way
North Adams, MA 01247
storey.com

Printed in China by Toppan Leefung Printing Ltd.
10 9 8 7 6 5 4 3 2 1

Library of Congress Cataloging-in-Publication Data

Names: Hyman, Frank, author.
Title: Hentopia : create a hassle-free habitat for happy chickens : 21 innovative projects / Frank Hyman.
Description: North Adams, MA : Storey Publishing, 2018. | Includes index.
  Identifiers: LCCN 2018012736 (print) | LCCN 2018015581 (ebook) | ISBN 9781612129952 (ebook) | ISBN 9781612129945 (pbk. : alk. paper)
Subjects: LCSH: Chickens—Housing.
Classification: LCC SF494.5 (ebook) | LCC SF494.5 .H96 2018 (print) | DDC 636.5/0831—dc23
LC record available at https://lccn.loc.gov/2018012736

This book is dedicated to Chris Crochetiere,
my first reader and muse. Your happiness is
my pole star. Thanks so much for wanting to
add chickens to our little urban homestead.

CONTENTS

Build feeders from free or low-cost materials or use vintage feeders. Create a critterproof "Vending Machine" feeder made from a 5-gallon bucket, eyehooks, and wine corks. Learn how to make a drawer feeder and a nursery pot feeder, and harvest grass clippings with your mower.

2. Drink Up (or Down)

Build a self-filling waterer from a rain gutter and downspout, a 5-gallon bucket, and watering nipples to keep chickens from "fowling" the water.

3. The Hen Pen and a Chicken Corral

Construct a predator-proof pen (aka "run") with readily available materials, and create a "Chunnel" (chicken tunnel).

4. Enrich and Accessorize

Entertain the chickens with "enrichments" (as zookeepers say) such as tree branches in the pen to roost on. Also use salvaged dog crates to isolate new, sick, or broody hens, and buy crickets at the bait shop to stage a rapid-paced rodeo for the chickens.

PART II:
INS AND OUTS
OF CHICKEN COOPS

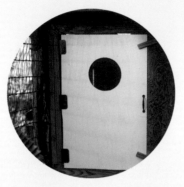

5. Doorways Four Ways

Build a no-sag gate for the pen, an automatic door for the chickens, and a henkeeper's door and cleanout doors for you.

6. No-Burnout Cleanout and Ventilation

Make coop cleanout super easy, as well as address the conundrum of how to have good ventilation without also having drafts. Covers bedding, easy composting (never smells and never needs turning), ventilation guidelines, and more.

7. Building the Best Nest Box

Make a nest box that is easy to clean and accessible even to kids.

Profile: Award-Winning Chef Suvir Saran's Heritage Breed Chickens

PART III: BACKYARD BUILDING 145

8. Tips on Tools for Non-Carpenters
146

Build your coop/pen/habitat with a jigsaw and drill. Learn about other safe-to-use tools and techniques.

9. High-Quality, Low-Cost Materials
159

Find free pallets, salvaged metal roofing, salvaged fencing, and the like, and learn how to use galvanized fencing, pressure-treated wood, cedar, or other options.

10. Coop Design
168

Find designs and how-to for a coop made of free pallets.

Profile: Author Amy Stewart's First-Lady Chickens 188

PREFACE
MY WIFE WANTED CHICKENS

I don't dislike chickens at all. I could live happily without them (although I would miss the eggs), but life doesn't always take you where you think you're headed. So when my wife said she wanted chickens, I knew we would have chickens.

Chris asked if the coop could have a pagoda roof. Unable to find pagoda designs on the Internet, I had to take my guerrilla carpentry skills to a new level.

A Bit about Our Experience

My wife, Chris, and I approached becoming chicken-keepers from two different angles. Chris, an animal lover and fervent cook, saw chickens as adorable pets with culinary benefits. My priorities were different. I wanted my wife to be happy (which I've learned is the key to being a happy husband), but I didn't want to be the guy who came home tired from a hard day at work and sat down with my beer just in time to remember some nasty chicken chore that I really should have done yesterday. A picture like that will put a fire under you. It sure did me.

Another issue is that I grew up working class with a frugal bent. I wondered how far down I could bring the cost of making a coop and keeping backyard chickens. Could the construction details be simple enough for people with limited skills?

Finally, I wanted us still to be able to take long vacations without coming home to starving chickens. In other words, I wanted to possess the least expensive and lowest-maintenance chicken compound in world history. A Hentopia for chickens *and* chicken-keepers.

With these goals in mind, we visited numerous chicken-keepers, thanks to the folks behind the Raleigh Tour de Coop, the Bull City Coop Tour, and the Greensboro Coop Loop (all in North Carolina). We scoured the Internet and flipped through books on coops and chickens. We took mental notes, written notes, and smartphone photos. We went back and forth on a few things, and we plowed through lots of graph paper.

Our coop and habitat design was more than a year in the making. We wanted something that would not only keep the chickens safe but also be easy to clean and convenient to maintain. If it allowed us to go away for two-week vacations without worrying about our hens, even better.

Our choices have paid off over the years. Our three middle-aged hens (a Buff Orpington, a Rhode Island Red, and a Black Marans) and four young hens (an Easter Egger, a Golden-Laced Wyandotte, and two Dominiques) lay about two dozen eggs a week. We spend time with them when we want to, not because we have to. Aside from periodically tossing them kitchen scraps and gathering their eggs, we can go for days without tending to the chickens when we're busy with other things in our lives. I can relax when I come home from work. I also get breakfast-for-dinner as often as I like (almost). The tagline for my twice-yearly coop design/build workshop is "We spend less time doing chicken chores than we spend cooking eggs."

We *have* been able to go away for as long as two weeks or more without a hitch. We've created a Hentopia. And you can, too.

If it allowed us to go away for two-week vacations without worrying about our hens, even better.

A bird's-eye view of our birds' domain

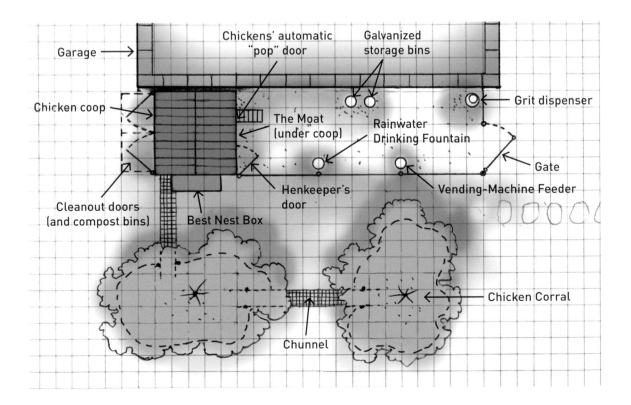

Garage

Chicken coop

Cleanout doors (and compost bins)

Best Nest Box

Chickens' automatic "pop" door

Galvanized storage bins

The Moat (under coop)

Rainwater Drinking Fountain

Henkeeper's door

Grit dispenser

Gate

Vending-Machine Feeder

Chicken Corral

Chunnel

A Bit about This Book

It's no secret that urban homesteaders, homeschoolers, hipsters, and handy folk of all types, as well as the not so handy, are acquiring backyard chickens. It's a certifiable trend, with understandable appeal. The birds offer many benefits:

1. Eggs with thick whites and orangey yolks

2. The Oscar-worthy entertainment value of birdy behavior

3. A reason to spend time outdoors

4. Insect and tick control

5. Weed management

6. Sustainable disposal for garden and kitchen scraps

7. Premium garden-fertility enhancement

8. Life lessons and responsibilities for kids

9. Biology lessons for kids (and adults)

10. Lessons in mortality for kids (and adults)

11. Small-scale business opportunities

12. The calming effect of contented clucking

One secret hasn't surfaced in the media yet: the countertrend of backyard henkeepers giving away their birds. At some point people discover that many of the common practices and techniques of henkeeping are time-consuming, expensive, inconvenient, or all three. Conventional coops are hard to clean out. Standard feeders and waterers are constantly dirty or empty or both. Predators penetrate flimsy pens to eat someone's pet-with-benefits. The list goes on.

Others who might enjoy the benefits are intimidated by the apparently high cost of conventional chicken-keeping.

I find animals fascinating and entertaining. I enjoy keeping them happy and well. But I don't enjoy it when their caretaking becomes a grinding chore. I suspect you feel the same way. Those unpleasant issues are a big part of the reason I've written this book on trouble-free and low-cost coops and sustainable hen habitats.

It's important to recognize that the primary beneficiary of a good design is the henkeeper, not the hen.

It's important to recognize that the primary beneficiary of a good design is the henkeeper, not the hen. Having kept dogs, cats, and fish and having worked with cattle, free-ranging rhesus monkeys, and loggerhead sea turtles, I feel that I have gained some sense of how critters see the world. I appreciate that animals have a social and emotional life. Nonetheless, I'm cautious about anthropomorphizing them — meaning that I refrain from seeing them as little humans with social and emotional lives identical to ours. This leads me to a point that I think is critical in designing a sustainable coop and hen habitat: You will benefit from that good design even more than your chickens will.

For example, whether their feed container is metal or plastic matters little to them, but it could make a big difference to you in terms of cost and effort. A key component of good coop and habitat design is to recognize that chickens will accept a wide range of options. That allows you to make a choice that fits your time, temperament, and budget. That's how you make your own backyard Hentopia.

Making the chicken-keeper's needs the priority doesn't mean the chickens will be neglected, however; in fact, it will be the opposite. If you set up a personal Hentopia that keeps you from feeling burdened, that means your chickens will enjoy a long collaboration with you. But if chicken-keeping becomes a chore, it can only lead to a downward spiral of neglected chores, chicken health problems, blown budgets, and ultimately an end to what had been a promising and tasty chicken-and-egg adventure.

In winter, the egg laying slows down just as the holiday baking picks up. A yellow store-bought yolk sets off the beauty of an orange homegrown yolk.

Recycled and Salvaged Materials

There often can be beauty and value in materials that other people have cast off, whether 100-year-old brick paving stones or a vintage copper weathervane chicken. Moreover, these items are often cheap or even free. I use these whenever I can, and you will find them in the photographs throughout this book.

Our Goals

In our design and construction process, my wife and I wanted our coop and hen habitat to have the following qualities, in no particular order:

- The coop and habitat look good.

- The birds are safe from predators.

- The areas don't need cleaning often.

- It is easy to clean out waste and put it in the compost.

- Feeding and watering the flock is easy.

- The feeders and waterers stay clean.

- The pen isn't muddy, but it does give hens a place to scratch for bugs.

- Foraging places are available in the yard but don't disrupt the gardens.

- It is easy to collect the eggs.

- The coop has good ventilation yet is weatherproof.

- Everything is built with common tools and low-cost materials.

- The whole setup is capable of being left safely on autopilot while we go away for two weeks.

These goals outline both the Hentopia that's in our backyard and the one that's in your own power to build. My practices in this book are by no means the only way to do things, but they will certainly save you time, money, and hassle compared to many other henkeeping practices out there now.

The bottom line is that keeping backyard chickens for eggs and companionship needn't be a chore. It can be a joy for henkeepers of all ages. Come along with us down the path to Hentopia.

The bottom line is that keeping backyard chickens for eggs and companionship needn't be a chore.

Domenica forages for tasty bugs and plants just as her jungle fowl ancestors did millennia ago.

PART I
CHICKEN HABITAT

Everyone talks about chickens needing a coop, but a coop is really just the box that chickens sleep in at night. Some of the trickiest parts of chicken-keeping revolve around all the things outside that box, which I call their habitat: feeders, waterers, the pen, and more.

So let's start with the chicken habitat. This section offers some short, doable, inexpensive innovations that will save you a lot of time compared to standard chicken-keeping practices. Building your chicken habitat will also help you get your skills up to speed for the bigger project of building that box for your chickens to sleep in.

1 Feed the Masses without Any Messes

Spending some time every day with your chickens should be a pleasure, not a prison sentence. Even the best henkeepers will enjoy having some time away from the coop — especially if that time away is spelled *r-o-a-d t-r-i-p*. The options outlined in this chapter allow chicken-keepers to enjoy two-week vacations without coming home to filthy feeders or a famished flock.

Let Gravity Do the Work

Poppy operates her custom-built vending machine.

Laying hens need to eat about a quarter of a pound of feed every day, varying of course with their age, activity, the number of eggs they are laying, the amount of free-ranging they do, and the season. The colder it is, the more they eat to keep their body temperature up. As they get older, they lay fewer eggs and need fewer calories. These kinds of factors aside, though, ¼ pound per day, per bird, is a good year-round average for figuring out how much feed to buy.

For example, our three older hens are just past the prime of their egg-laying lives, yet they are still productive. The four young hens are at their peak of egg laying. For us, a 50-pound bag of organic layer feed lasts about a month in warm weather and a little less in cold.

The important thing is that we don't have to feed them very often. We use a vintage galvanized gravity feeder we bought for $5 at a junk-tique shop. It's big enough to hold about 25 pounds of feed, so I have to top it up only every couple of weeks. The remainder of the bag stays in a small watertight and verminproof galvanized garbage can next to the coop.

If you have many more birds than this, you can set up multiple gravity feeders.

Use Nursery Pots as Gravity Feeders

My friend Katie Ford had been feeding her two dozen chickens with three small gravity feeders that needed topping up every couple of days. When I showed up one day with a stack of 25-gallon black plastic nursery pots, she sized them up right away as ready-made gravity feeders and said, "I need one of those." She quickly put one in the run and dumped in a 25-pound bag of pellets. The chickens went right to work.

Katie realized the new feeder would hold enough feed for a week or two. That's a big time-saver for a first-grade teacher trying to expand her backyard homestead.

These nursery pots have a lot of advantages for chicken-keepers who want to use them as gravity feeders:

- They're ready-made to spill out chicken feed continuously from day one, thanks to several drainage holes distributed along the bottom edge.

- They're cheap, and often free if you ask your local landscaper or nursery for a few. Or scoop them up when you see a landscape crew doing a planting job. They'll be happy for you to take them off their hands.

- They're durable. Most plastic objects left in the sun will break down into tiny pieces that blow away and get into everything, but nursery pots have ultraviolet light (UV) inhibitors in their plastic, which keeps them from breaking down. They can stand being in full sun. Nurseries re-use them for many years without loss.

- They come in a range of sizes that can match your project needs: 1-gallon, 3-gallon, 5-gallon, 10-gallon, 15-gallon, and 25-gallon are some of the most common sizes. Some are much bigger than any gravity feeder on the market; small ones could feed chicks inside the brooder.

- They're easy to handle. The bigger pots have handles on the rim, making them easy to move even when filled.

- They don't uglify the garden. They're black, which goes with any-thing! Aesthetics may not be your most important consideration, but a good rule is that any object in the garden that isn't gorgeous should be a dark color so that your eye is more naturally drawn to the brighter, more attractive elements in your yard: plants, garden accents, and of course lovely chickens.

Katie positioned the pot in the run on her feeding station: a pallet set on bricks, where the three small feeders had been parked. The bricks keep the pallet off the ground so it won't rot. With the feeders raised off the ground, the chickens can't scratch debris into them. A piece of metal roofing mounted overhead keeps the rain off the feeders and the pallet.

Being handy, Katie immediately went to work customizing her new gravity feeder. Realizing that the drainage holes needed to be bigger to allow more pellets to flow, she took a knife and widened each hole to about twice its original size. Though the plastic is sturdy, it easily yields to most knives, or even pruners. Next, she cut down an old baby pool to function as a saucer to hold the pellets.

And then she went to place an order for more feed.

Park the nursery pot under a roof and on a dry surface onto which the feed pellets can gently spill out.

The drainage holes are about one inch wide. Use a knife to cut them to about two inches wide.

Gravity will keep pellets coming down through the holes.

Mounting a gravity feeder on a grain scoop with some stiff wire or a couple of screws is an easy improvement that saves a lot of time and hassle.

Design a Bottomless Food Drawer

A gravity feeder is, basically, a metal tube mounted over a metal saucer. As the chickens eat from the saucer, gravity pulls more feed from the tube. The saucer refills automatically until the tube is empty. Who needs robots when you've got gravity?

The three challenges with gravity feeders are:

1. Hens scratching debris into the saucer

2. The hens' tendency to get on top of the feeder and poop on their feed

3. The hazard of rain ruining the feed

We like having a run that's open to the weather, so we can't keep a feeder in the pen; it would get soaked. We solved the poop and rain problems by parking our 18-inch-high gravity feeder under the coop in the space I call the Moat. The Moat, which is a barrier for vermin, also creates shelter from the rain for the hens and the feeder. And without enough headroom to roost on the feeder, they can't poop in it either.

To keep a chicken from scratching debris into the feeder, it's smart to raise the saucer up to chest-high on the birds. We do that with a couple of scrap pieces of 4×4 lumber.

But that setup can introduce a fourth challenge. Once it's empty, how do you retrieve, refill, and return the feeder to its spot under the coop without skinning your knees, straining your back, or banging your head on the coop?

As I puzzled out this design issue, I felt as though I needed something like a kitchen drawer, so I could pull the feeder out from under the coop and then slide it back into place without bending over. In my garage I found a short-handled grain scoop I had bought at a yard sale for a couple of bucks. I parked the gravity feeder on the blade of the grain scoop, fastened them together into one unit, and I was in business. No knee-skinning, back-straining, or head-banging required. The hens' food stayed dry, poop-free, and accessible for weeks — while we went on vacation.

The hens' food stayed dry, poop-free, and accessible for weeks while we went on vacation.

Storing a gravity feeder under the coop keeps out rain and poop. Attaching the feeder to the grain scoop makes it easy to retrieve, refill, and replace.

HOW TO MAKE IT

EASY-ACCESS FOOD DRAWER

When the food drawer is properly in place, only the end of the handle protrudes from under the coop. When the saucer looks low on food, you can pull the food drawer out and refill it, then slide it back in. A nursery pot (page 22) would also work as a gravity feeder mounted on a scoop.

1. Turn the feeder upside down on the ground and lay the shovel blade upside down on the saucer. Drill a hole through the approximate center of the blade and through the saucer. The hole should be big enough for a ³⁄₁₆"-wide, 2"-long bolt to slide through it.

2. Turn the scoop and saucer together, so that you can insert the bolt through the hole from below with one hand and with your other hand screw on the matching washer and nut as snugly as possible.

3. Turn everything right-side-up and fill the feeder.

4. Slide the blade under the coop or other rain cover and onto the tops of two 4 × 4s placed parallel as if they were a set of railroad tracks.

Build a Critterproof Vending Machine for Your Flock

Violetta enjoys her reward for giving the corks a good whack. The eyebolt acts like a toggle switch and drops pellets from the bucket.

A good chicken feeder should work smoothly every day, like a well-stocked vending machine. It should also be easy to keep clean, should keep the food dry and should be off-limits to feral critters that refuse to pay their rent with fresh eggs.

For four years we had been happily using a gravity feeder. But marauding gangs of sparrows had started plundering the feeder's open saucer occasionally. Despite their big appetites, sparrows are small enough to fit through the 2" × 4" openings in the fencing. Thank goodness mourning doves can't get in, or I'd have been buying feed twice a day. We haven't had signs of rodents, but I was sure some would show up eventually for the grainy buffet. Coming up with a new and improved critterproof feeder was on my to-do list.

Four-inch-long eyebolts with ¼"-diameter threads make a good toggle switch for your chickens' vending machine. Drill ⅝" holes in the bottom of the bucket and a 3/16" hole in the corks for a durable, low-tech device.

Some chicken-keepers choose to keep sparrows and rodents away from their feeder by enclosing the entire pen in hardware cloth. This isn't a cloth at all, but rather a type of heavy-gauge wire fencing with ½" × ½" openings. It is heavier, costlier, and more unwieldy than the 14-gauge fence wire commonly used for enclosing hen pens. For example, a 3' × 50' roll of hardware cloth weighs half again as much and costs three times as much as the same size roll of 14-gauge wire fencing. So I ruled that out.

There are also some boxy, verminproof feeders made of either metal or cedar available online. The chickens step on a treadle that opens the lid on the feeder, allowing them to eat. When the chickens walk away, the treadle pops up and the lid pops down, closing the feeder and blocking any songbirds or rodents. These treadle feeders cost $200 or more. There are construction plans available online for the wooden ones, but making the angled cuts and putting together the lever arms would be time-consuming and tricky. Plus, if enough debris gets under the treadle to block it, the feeder won't open and the chickens go to bed without their supper. So I ruled out that option, too.

Thus, I was very glad to find a video by "Coach G" detailing how to build a vending machine (see References and Resources, page 198). I've customized it a bit for our Hentopia coop, using corks instead of duct tape or glue. The vending machine holds almost a full bag, so we still fill it only about once a month for our seven birds.

In this device, a 5-gallon bucket (with a tight-fitting lid to keep out the weather and critters) hangs from a post at a height that allows the hens to peck at a couple of corks screwed onto eyebolts run through holes in the bottom. As the eyebolts move — like a toggle switch — small amounts of feed spill out of the holes. Hens eat the feed, leaving little or nothing for freeloading vermin.

CONSTRUCTION TIPS

By drilling a hole just the diameter of the shaft of the hook you're going to use for hanging, you allow room for the shaft to enter the wood without splitting it, while the threads slice their way into the wood. The threads hold screws in place, making them superior to nails.

If it's too hard to screw the hook in by hand, here are two things you can do: First, rake the threads lightly across a bar of soap. The little bits of soap will act like a lubricant, making it easier for the screw to go in. Second, you can insert a screwdriver into the hook (as shown on page 30) to gain the leverage to twist it into place.

HOW TO MAKE IT

CRITTERPROOF VENDING MACHINE

WHAT YOU NEED

- 5-gallon bucket with tight-fitting lid
- Drill with a ⅝" spade bit and a ³⁄₁₆" bit
- Two wine corks
- Pair of pliers
- Two 4"-long eyebolts with ¼"-diameter threads
- 4"-long hook with a threaded shaft

WHAT TO DO

1. Drill holes in the bucket.

Using the ⅝" spade bit, drill two holes in the bottom of the bucket. The eyebolts' shafts will hang down through these. Put each hole roughly below the spots where the bucket handle attaches (the reason for this will be clear later).

2. Prepare the corks.

Using the ³⁄₁₆" bit, drill into the bottom of each wine cork a little more than half the distance from one end to the other. This size hole makes room for the eyebolt shaft to enter without splitting the cork. Hold the cork with a pair of pliers so you don't hurt your hand if the bit goes awry. (In the photo, I'm using a champagne cork for a more celebratory look.)

3. Attach the corks.

With one hand inside the bucket holding the eye of one of the eyebolts and the other hand outside the bucket holding one of the corks, screw the cork onto the shaft of the eyebolt. Repeat with the other eyebolt and cork.

4. Select a spot to hang the bucket.

With the eyebolts and corks in place, choose a spot in the pen or coop from which to hang the bucket. You want the corks at the height of the chickens' heads. Mark the spot for the hook.

5. Screw the hook in place.

As with the corks, use a drill bit that matches the diameter of the hook's shaft so that there is room for the shaft to enter the drilled hole while allowing the threads to bite into the wood. Screw the hook into place.

6. Add feed, hang the bucket, and observe the feast.

Fill the bucket with feed and put the watertight lid on. Hang the bucket from the hook. Tap the corks so that some feed drops out of the bucket. Now you see why the position of the holes matters. If there's a hole at the back of the bucket, much of the feed will bounce outside the fencing.

Making It Work

Remove other sources of food to focus the chickens' attention on their new vending machine. Depending on how smart your chickens are, they will learn in a few minutes — or, like ours, a couple of days — how to trigger the corky toggle switches. Some will get it right away from watching you. With others, you may need to grab them and tap their beaks against the cork until you see them get the idea. Some henkeepers have had luck shining a red laser light on the toggles to get the hens to peck, but ours weren't interested. When they got hungry enough, tapping on the corks started making sense to them.

Now they seem to really enjoy tapping the cork on their new vending machine. And the marauding sparrows have disappeared, too.

Round Out Your Chicken Feed

At a chicken-keeping workshop I heard a speaker say that a chicken's diet should be about one-third grains, one-third greens, and one-third bugs. As you might guess, the speaker was selling a system for growing black soldier fly larvae as chicken feed. I was willing to give it a try. Sounded cool. Lots of chicken-keepers rave about soldier fly larvae as a food source. Yet, strangely, Chris has vetoed any backyard projects containing the word "fly."

But who doesn't love cuddly, wiggly worms? Growing red worms has not been vetoed.

An In-Ground Worm Bin

The topic of growing worms, called vermicomposting, is big enough to require a book in itself. (See References and Resources for more on vermicomposting.)

I'm developing an in-ground worm bin that's 4 feet by 8 feet across and 1 foot deep. It's functioning pretty well, but I'm still working out some of the kinks. I'm sure that in another year it will be a smoothly operating red-worm chicken-food-propagating machine. The large size means I can grow large populations of red worms. Its being in the ground means the soil will moderate the summer and winter temperatures that would otherwise inhibit the worms' growth. During dry spells I maintain the proper moisture level by slowly draining a rain barrel into a soaker hose over the worm bin.

Our chickens and regular compost bins already do a great job disposing of kitchen scraps and most garden scraps, so the worms' diet will depend on other free sources of organic matter: chaff from local coffee roasters, woody plant scraps, newspapers, cardboard, and irreparable cotton clothes.

Keep in mind that the worms cost about $25 a pound plus shipping, and they are easy to kill by accident or negligence. So take my outline as a bit of inspiration and then do the smart thing: get a book or take a class on vermicomposting or growing soldier fly larvae or mealworms for chicken feed.

We've also developed a smaller source to supplement the hens' insect diet. We use sticky traps to catch the pantry moths that occasionally show up in the kitchen and the cave crickets that multiply in the crawl space. When the sticky traps are full, we set them in the pen.

Even three birds scratching and foraging can wreck a lawn in a week or two.

POULTRY POINTS
Safe Shells

Leftover eggshells are full of calcium, and laying hens have a strong appetite for calcium. It's safe and widely recommended to feed leftover shells to hens. Do smash them flat, though, so the birds don't recognize them as eggs. You don't want to give them any ideas about eating their own eggs.

Chickens love these treats, dead or alive. Surprisingly, their beaks do not stick to the traps. The sticky traps are cheaper and safer than pesticides, and the organic, unpoisoned six-legged vermin make our hens happy and save us a tiny bit on chicken feed.

As for the grain portion of our chickens' diet, that's easy. We've found local distributors for organic layer feed. Supporting organic farmers means fewer chemical fertilizers, herbicides, and pesticides getting into our nation's drinking-water supplies. Organic chicken feed costs more than nonorganic, but that's largely because there are so few feed mills dedicated to producing organic grains. Since organic mills are fewer and farther between, a lot of the higher prices we're paying are the freight charges. But more mills are being established, so prices should fall in the next few years.

Greens for the Girls

As for the grassy portion of our hens' diet, we had to cut them off from free-ranging in our tiny yard. As you may know from experience, even a few birds scratching and foraging can wreck a lawn in a week or two. Let's not even talk about the rapid distribution of chicken fertility all over the deck and patio. Instead, I have found some other tools and techniques for keeping more greens in the girls' diet:

- Naturally, we give them leaves of garden greens that have bolted.

- Developing a friendly rapport with the staff at the grocery store means we can bring home cabbage leaves and other nonsalable food scraps.

- Almost any weed pulled from our garden beds will generate a feeding frenzy. Common chickweed is a favorite. No idea where that name came from.

Pigweed and other weeds (such as chickweed, henbit, annual fescue, crabgrass, and purslane) are valuable additions to your hens' diet.

I've also been taking advantage of some biennials in the vegetable garden to provide greens for the girls. We always get a great crop of carrots, and from each harvested carrot we toss the tops in the hen pen. Chickens will knock each other down trying to get to them first. Some carrots get too big and woody for us to eat, so I pull those from the vegetable garden and replant them next to the pen. Since they are biennials, they keep growing and will resprout tops that we harvest for the chickens.

If I wanted to spend my time and energy carrying chickens back and forth, I could deposit them in a vegetable bed at the end of its season. Then I'd have to fence in that bed to keep them out of other beds growing crops we want to eat. If you want your flock to forage in the garden, wrap suitable beds with chicken wire and corner posts to keep the birds out of trouble. Make sure the chicken wire is high enough to keep them from flying out. You may also be able to channel your birds from the pen to the vegetable beds with a "chunnel." More about those in chapter 3.

About the time Chris vetoed the fly project, I belatedly realized I was already harvesting greens from our yard every week when I mowed the lawn. I just needed to get these first-rate greens to the chickens.

I dug through piles of might-be-useful-someday objects in the back of the garage, looking for the bagging attachment for our mower. It had never seen any action because our mower shreds the clippings and lets them fall back into the lawn to feed the earthworms and replenish the fertility of the grass. You need little or no lawn fertilizer if you don't take away the grass clippings.

Now, once a week, the girls get a bag of fresh grass clippings to peck at and devour. I figure the lawn can spare one bagful.

Our mower, a cordless electric, is one of my favorite power tools. Here's why:

- It never runs out of gas or oil.

- It has no spark plugs or filters to be cleaned or replaced.

- It's very quiet, no louder than a box fan.

- It starts with the push of a button rather than countless yanks of a cord. The battery charge lasts for about 45 minutes, which is more than enough for our small yard as well as our neighbor's (with whom we share the mower). There are brands with removable batteries; you can keep a second battery charged and ready to swap out if you have an oceanic lawn.

A third of our hens' diet isn't yet from greens or bugs, but we're getting there. Maybe I should start growing duckweed in the water garden. Or set a Japanese beetle trap in the pen.

> **I belatedly realized I was already harvesting greens from our yard every week when I mowed the lawn.**

2 Drink Up (or Down)

Water, water everywhere, but not a drop that's fit for a hen to drink. If the water is in a saucer, the girls have mucked it up by scratching dirt into it. If it's in a gravity-fed waterer, they've hopped on top and deposited some of their natural fertilizer into it. (Yuck.) And the only thing worse than dirty water is no water at all, when they've knocked over the waterer or drained it on a hot day.

There is a better way. We sustain our hens with a simple system that keeps the waterer clean, free of debris, and topped up almost year-round.

Our Dominique pullet, Domenica, quickly learned to tap the nipples to get fresh water.

Bring Rainwater to Your Birds

My design professor at North Carolina State University, Will Hooker, was an early champion of backyard chickens. He captured my attention with his system for using free and plentiful rainwater for his chickens. He screwed a store-bought rain gutter to the eave of his coop's roof to capture rainwater. The water drops through a downspout into a 35-gallon plastic cistern with a tap. Will uses this rainwater to refill his chickens' conventional gravity-fed waterers.

My wife and I saw Will's setup long before we got serious about chickens. I knew even then that we would borrow — I mean, steal — this idea of harvesting rainwater if we had hens.

Once we got a little more serious about henkeeping, I found two websites by homesteader Anna Hess: The Walden Effect and Avian Aqua Miser (see References and Resources). She describes how she and her husband make (and sell) hanging poop-free waterers consisting of a 5-gallon bucket with livestock watering nipples protruding from the bottom.

By combining these two techniques, we have a chicken watering system that rarely needs our attention. There's no saucer to get dirty or tip over. A repurposed metal hanging basket crowns the bucket to keep hens from roosting and pooping on the waterer. (A bucket lid with an opening for the downspout would do the same.)

Since the East Coast averages a good rain every 10 days or so, we rarely have to top up the waterer, nature does it for us. We've only refilled it ourselves maybe a dozen times in four years, usually when we wash it out once a season. Obviously, in drier climates the chickens will be thirstier and the rains will be fewer, so use multiple containers or top it up by hand more often.

How It Works

Some people attach nipples to PVC pipe (fed by a waterer) instead of the bottom of a bucket, and that may work for your pen. But the curved wall of the PVC pipe may make the nipples more likely to leak than the flat bottom of a bucket.

How many nipples will you need? Allot one nipple for every three birds. One bucket can easily sport three nipples. If you have more than nine birds or you find your chickens queued up in long lines waiting their turn to drink, then add another bucket.

Bottom line: Chickens need to drink about a half-pint of water a day, the actual amount depending on their size and how hot it is. A 5-gallon bucket holds enough water — 40 pints — to last our three hens about 27 days without needing refilling or being spoiled by poultry pollution. This setup is one of the biggest contributors to our goal of enjoying fewer chicken chores and more worry-free vacations. So let nature and gravity do the work of keeping your hens watered. You've got places to go.

Project Notes

Installing a rain gutter and downspout, as in the following project, will cut your ongoing chicken chores dramatically. The gutter and downspout fill the waterer with rainwater from the coop's roof (or any other nearby roof).

When it rains, the roof edge will act like a waterfall, dropping the water into a 3- to 4-foot-long gutter. From the coop roof you'll probably need 6 feet, max, of downspout to reach from the end of the rain gutter to the top of the waterer. A 5-gallon bucket will fill up from just a few minutes of steady rain. Make sure the gutter runs slightly downhill toward the downspout so the water flows down and doesn't sit stagnant.

We have a chicken watering system that rarely needs our attention.

Tools and Materials

You can pick up gutter scraps and downspouts pretty cheaply at your local metal scrapyard, or call around to gutter companies and ask them to hold a few short pieces for you. You can also buy gutters and downspouts at a big-box home improvement store, or possibly at an architectural salvage store.

Use self-tapping screws and a matching driver bit from a hardware store so you won't need to predrill any holes in the gutter or downspout. You'll also need an $^{11}/_{32}$" drill bit to make the holes in the bottom of your bucket. That will be just the right diameter for the threads of the nipple to bite into the plastic and create a watertight seal. Spend the extra money for the nipples that have a rubber gasket so that you'll be sure they won't leak.

Can't position your waterer directly under the gutter? A flexible downspout will solve that problem. If you choose to have a lid on your bucket (see page 42), mark the spots where the corners of the downspout opening meet the lid. Use a drill to make a hole at each corner, use a jigsaw to connect the holes, and pop the downspout in.

HOW TO MAKE IT

RAINWATER DRINKING FOUNTAIN

WHAT YOU NEED

- Tape measure and pencil

- 3- to 4-foot length of gutter

- Metal snips

- 6-foot length of downspout

- Self-tapping screws and matching driver bit

- Watering nipples, preferably with a rubber gasket: 1 for every 3 birds, up to 3 nipples per bucket

- 5-gallon bucket

- $^{11}/_{32}$" drill bit

- 4" zinc-coated screw-in hook (the type used for hanging flower baskets) or bucket holder (see References and Resources)

WHAT TO DO

Prepare gutters.

1. Use tape measure, pencil, and metal snips to mark gutters and cut them to length. Leave the end caps off to make it easier to clean out leaves from the gutter.

Set up the gutter and downspout.

2. Using the self-tapping screws and matching bit, screw the gutter to the ends of the rafters and just under the edge of the roof. Depending on the design of your coop, you may need to add scrap pieces of wood to the rafters or walls to position the gutter under the roof edge.

Attach the nipples to the bucket.

3. Use up to three livestock watering nipples for each 5-gallon bucket.

4. Drill holes in the bottom of the bucket. Use an $^{11}/_{32}$" bit to make a hole the threads will bite into.

5. Screw in the nipples. If the weather is very cold, the plastic of the bucket may be too hard for the threads to cut into. If so, let the bucket warm up indoors before installing the nipples.

6. Ready to fill and hang. Periodically check for leaks and the occasional defective nipple by pouring a few inches of water into the bucket and testing the nipples.

Mount the waterer.

7. For drilling a hole to mount the bucket hook, choose a bit that is the same diameter as the shaft of the hook, but narrower than the diameter of its threads.

8. Select a spot for the hook that will let the bottom of the bucket hang about 12 to 14 inches above the ground, so the chickens can reach the nipples.

You'll see the light dawn when your bird realizes where her water will come from.

LIDDED OR LIDLESS?

Some henkeepers let the downspout reach into the bucket by cutting a matching hole in the lid. That's one way to do things. The lid keeps out poop but doesn't preclude algae, mosquitoes, or leafy debris from entering the waterer; they get in through the downspout. A lid just makes it harder to see into the bucket.

I prefer my waterer without a lid so that I can see the water level and more easily add apple cider vinegar. For that reason, I plopped a repurposed upside-down hanging plant basket over the bucket as a decorative guard to keep hens from roosting there.

Our downspout stops just short of the bucket, so rainwater drops in just fine. The advantage of going lidless and with a shorter downspout is that it's very easy to remove the bucket whenever I want to clean it. With or without a lid, you'll need to clean buckets a couple of times a year to keep the nipples from clogging up due to algae or debris from the roof.

Making It Work

Your chickens won't instinctively know how to use the nipples to get water. Your best bet is to catch the dominant chicken with both hands around her wings and body and hold her beak under the bucket. Lightly bump her beak against a nipple until water dribbles out. At first she'll squawk, wondering why you're tormenting her so. But you'll see the light dawn when your bird realizes where her water will come from and she starts drinking.

If you're lucky, the other chickens will see the dominant hen drinking from the nipples and they will get the idea, too. Don't count on it. They may each need their own intro to the watering nipple. Still, that's so much easier than cleaning up dirty saucers of water. Be sure to remove other water sources, or they'll never switch to the new waterer.

The Freeze-Free Solution

One of the biggest responsibilities of a henkeeper is to keep our birds' water supply from freezing. It's bad enough that the chickens have to go without water when icebergs form, but it also means that we (and by "we" I mean Chris) have to head out into the blizzardy weather with a kettle of boiling water or a hammer to break up some ice.

For colder parts of the country, there are many months straight of subfreezing temps and rock-solid ice in the waterers. Fortunately for us in North Carolina, freezing is a problem only a few times a winter; rarely does the waterer get more than a top layer of ice. But even that kind of daily response won't work if we're on vacation. I had to find something better.

Some folks recommend wrapping electric heating tape around the base of the waterer. Since the plastic of the bucket is an insulator and not a conductor, however, that is reliable only into the upper teens. Plus, the setup is time-consuming for just a partial solution.

Other henkeepers recommend aquarium heaters. They are not expensive, and they're made to be in the water. But chatter in the online chicken chat rooms indicates that aquarium heaters wear out quickly. They are designed to push the heat up in a 60°F (20°C) room only to about 78°F (26°C). Outdoors, these heaters are in way lower temperatures, so they work long and hard until they fail. And when they fail, the ones made of glass burst and electrify the water and the metal part of the drippers. The voltage is low, but think of what that does to chicken lips.

At last, on the Backyard Chickens forum, I found the best option: birdbath heaters. Of course. They're made to work outdoors, in winter and underwater. It's a plug-and-play situation. Plug in a birdbath heater, and drop it in the bucket. If you have a lid for your bucket, you can leave it ajar for the season since neither mosquitoes nor algae are active in winter. The girls will be so glad to have warm, liquid water available all winter. Throw in a few tablespoons of apple cider vinegar, and they'll think they're having a nice afternoon tea.

A birdbath de-icer in a chicken waterer is perfect for winter conditions. You may also need an extension cord rated for outdoor use, a watertight cord connecter, a GFCI outlet, and a thermostat.

Watertight cord connector

Thermo Cube thermostat

Before you retire to the house for your own teatime, I highly recommend adding a few other devices for safety and comfort:

Extension cord and watertight connector. If the cord from your birdbath heater isn't long enough to reach the outlet, you'll need an extension cord. Buy one that's UL rated for outdoor use, and fit a watertight connector over the cord connection to prevent shorts.

GFCI outlet. There is a small chance that the cord or the heater could short out in wet weather. A short could then go past the outlet, heat up some wires, and cause a fire in your house or garage. To prevent this, plug the cord into a ground fault circuit interrupter (GFCI) outlet, which will turn itself off in response to a short and save your house. After you fix the short, pushing a button on the outlet revives it. A licensed electrician can switch any normal outlet to a GFCI one. It's worth it: you're dropping a live electrical device into water.

Thermostat. If your birdbath heater doesn't come with an internal thermostat, you'll waste a lot of energy if the weather fluctuates. A Thermo Cube TC-3 device has a simple internal thermostat that lets current flow only when the temperature drops below 35°F (2°C) and turns off when the temp rises above 45°F (7°C). In no time, the TC-3 will pay for itself by saving energy and extending the life of the heater.

HOW TO MAKE IT

DE-ICED TEA FOR THE GIRLS

WHAT YOU NEED

- Birdbath de-icer
- Extension cord rated for outdoor use
- Watertight cord connecter
- GFCI outlet
- Thermo Cube TC-3 thermostat

WHAT TO DO

1. Insert the de-icer's plug into one half of the watertight cord connector. Thread the extension cord through the other half.

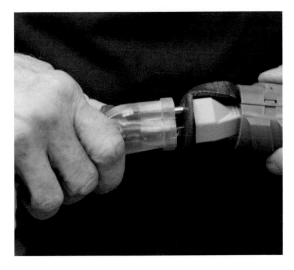

2. Plug the de-icer into the extension cord. Be sure to slip the included gasket over the metal plug prongs for extra protection against water.

3. Screw the two halves of the watertight cord connector together.

4. Place the de-icer in the water bucket. If your bucket has a lid you can put it on loosely. In winter, algae and mosquitoes won't be a problem.

Note: If your de-icer lacks a built-in thermostat, plug the extension cord into a Thermo Cube thermostat. Then connect the entire works to a GFCI outlet for safety's sake.

3 The Hen Pen and a Chicken Corral

Almost every critter likes to eat chickens, and that includes our sweet dogs and cats, both domesticated and feral. For most of them, a running chicken triggers an instinct to hunt the way a matador's cape calls a bull to charge. As your chickens' security manager, you'll want to make the right fencing choices to keep the carnivores in a constant state of disappointment. Read on for my tips on the tools, techniques, and materials that get the job done.

Setting the Posts for Your Hen Pen

I'm particular about setting posts. I've seen fences where the posts are out of line and leaning all helter-skelter. These raggedy fence lines could have been prevented with judicious use of a taut string, a plumb stick, and concrete in the posthole. This first section describes how to set posts for your hen pen that won't embarrass you in years to come. (Note that I use the terms "run," pen," and "hen pen" interchangeably.)

Predators love chicken. A proper fence sends them away hungry.

The 14-gauge welded wire on this pen will keep out common predators, including domestic dogs.

Post-setting tools: from back, plastic tub, wheelbarrow, water bucket, rubber boots, posthole digger, sharpshooter shovel, 1" × 2" stick, hoe, concrete mix, corner level, standard level

Tools and Materials

Buy pressure-treated posts, which should last as long as 40 years, from your local building materials supplier. They can be 4" × 4" or 6" × 6" thick (you'll need to widen the posthole a bit to accommodate the larger posts) and at least 8 feet long. With 18" in the ground that leaves 6.5 feet above ground, enough clearance for a tall person to walk in the pen.

The term pressure-treated means that the wood has been bundled together and steeped in chemicals that repel rot and termites. The atmospheric pressure in the treatment chamber is boosted so that the chemicals penetrate the wood, ultimately mimicking the natural preservative that redwood, cedar, cypress, and other rot-resistant woods already have in their heartwood.

In the old days the chemicals included highly toxic products like arsenic and chromium. Fortunately, the Environmental Protection Agency hasn't allowed the use of arsenic and chromium in pressure-treated wood since 2003. Instead, you can now buy what's called alkaline copper quaternary (ACQ) pressure-treated lumber at any construction supply house. For its fungicide, ACQ-treated wood simply has lots of copper, the same safe material that your water pipes are made of and that occurs naturally in the soil. ACQ treatment also stops termites with a mild chemical called quaternary ammonium, which is used as a disinfectant in kitchen products like Formula 409 spray cleaner. You wouldn't enjoy drinking it, but what little might leach into the soil is very far down the list from the heights of toxicity that arsenic and chromium occupy.

TOOL TIPS

- Using paint or tape, make a mark on the handle of your **posthole digger** so that you can see when the hole is about 18 inches deep.

- You can save a little time by using a **corner level**. It will tell you if your post is plumb (perfectly vertical) in the north-south direction and the east-west direction at the same time. When you're done, the gadget folds up for convenient storage.

- For mixing the concrete, a garden hoe will work, but a **concrete-mixing hoe** with a couple of holes in the blade is moderately better.

- All of my plastic water buckets eventually cracked, so now I use a **2-gallon metal bucket** on which I've drawn marks to show the number of quarts it holds.

- To drop the concrete into the hole, a narrow-bladed transplanting shovel, called a **sharpshooter**, works best. Use a 1" by 2" stick about 3 feet long to pack the concrete down into the hole.

The Beauty Part

We can see our hen habitat from our back door and our screen porch. To make it more attractive, I used the same type of wood — unpeeled cedar posts — for both the pen posts and the corner posts for the coop itself. Since the coop was going to be 5 feet across the front, I also set all the posts for the pen on a 5-foot spacing. This line of identical posts set with a rhythmic spacing between them ties the two structures together visually. With all the posts rocked in with concrete, I don't have to worry about them getting out of kilter, either.

Your coop and pen don't have to use the same type of posts or the same spacing. That's just one option. A common spacing for posts for a pen is eight feet.

For novices, it is easiest to install one section of wire at a time from post to post. Otherwise, someone has to hold up an entire, heavy roll of fence wire.

HOW TO MAKE IT

POSTS FOR THE PEN

Marking and digging the postholes can be a one-person job. So can mixing the concrete and setting the posts. But it goes a lot faster and easier with two people doing the work.

WHAT YOU NEED

- Pressure-treated posts, 4" × 4" or 6" × 6", at least 8 feet long

- Posthole digger

- Plastic or metal tray or tub to mix concrete in (or, in a pinch, a wheelbarrow)

- Concrete mix (e.g., Quikrete or Sakrete)

- Shovel (I recommend a sharpshooter shovel) for dropping concrete into posthole

- Bucket with quart levels marked to show how much water to add to dry concrete in mixing tray

- Concrete-mixing hoe (or a garden hoe)

- Two levels (also called plumb sticks), or a corner level

- Skinny, 3-foot-long board to pack concrete with (a 1 × 2, a 2 × 2, or even 2 × 4)

WHAT TO DO
Plan and dig the postholes.

1. Figure out where your corner posts go, and dig holes for them with the posthole digger. You can drop the dirt from the postholes onto a tarp or into a wheelbarrow to make it easier to dispose of.

3. After reading and fully understanding the mixing directions, cut the bag across the middle with the shovel.

Prepare the concrete.

2. Put the mixing tray a few feet from the posthole and drop an 80-pound bag of concrete mix into the tray. Sometimes you can find 40- or 60-pound bags; they are much easier to handle. The directions on the package will tell you the right amount of water to add.

4. Flip the bag over and empty it out as if it were an eggshell you've cracked on one side and are opening to let the yolk and whites fall out.

5. As if you were mixing dough, gently pour the water into a depression in the middle of the concrete. Pour in only half of the recommended water; after mixing this in, using the hoe, gently add the rest and mix again.

6. Mix the concrete according to the directions until it's thick and moist all the way through like bread dough. If it's dry and crumbly or wet and runny, it won't set hard enough to hold the post upright over time. You can always add more water, but you can't take any out, so add small amounts of water as needed.

Set the posts and pour the concrete.

7. Set a post in the hole, resting it on hard ground. Hold a level against the post on two adjacent sides, or use a corner level, to get it vertical in two directions: north-south and east-west. That's called getting the post "plumb."

8. With the post person holding the post plumb, the concrete person can scoop up concrete with the sharpshooter shovel and carefully drop it in on all four sides of the post.

9. As the concrete fills the gaps in the post-hole, the post person keeps the post in place with one hand and with the other pushes the concrete down with the skinny board, packing it firmly and eliminating air pockets.

10. Fill the hole until the concrete is just above soil level and smooth it off so that any rain will flow away from the post. The moist concrete will hold the post upright as it sets; there's no need for attaching supports to the post.

CONSTRUCTION TIP

After you've dug your two corner postholes and set those posts, run a tight string between them about ankle-high so that you can mark the intermediate postholes and get them exactly in line with the corner posts.

Stella d'Oro keeps watch. Raccoons, possums, coyotes, and weasels are common predators, but domesticated dogs are probably a bigger threat to backyard hens.

The Best Defense Is a Wire Fence

Now that the posts are placed and ready, it's time to get them wired: welcome to Fencing School. We'll begin with the best choices to build your Hentopia.

Tools and Materials

For the most durable and cost-effective *walls* of a chicken pen, you'll need what country folks call "hog wire" and city folks call "dog wire." In hardware stores it's called "14-gauge galvanized welded wire fencing" and has openings that are 2 inches by 4 inches. This fencing comes in rolls with a height of 2, 4, or 6 feet.

Chicken wire, with its hexagonal patterns, may seem to the novice like the obvious choice for dissuading wily predators. It is great for protecting garden beds from free-range birds and for keeping new birds and sick birds quarantined. But when some urban henkeepers I knew used chicken wire for their pen, a motivated pack of stray dogs had no trouble pulling the hexagons apart with their teeth and breaking my friends' hearts.

Though it's aptly named for keeping chickens in, chicken wire is no good against four-footed predators. Its main value is as an inexpensive *roof* over the pen: it will protect your birds from hawks and other raptors.

On the Fence

Before going further, let's learn some fencing terminology.

Gauge. This term refers to the thickness of the wire. Chicken wire, for instance, is 20-gauge, which sounds pretty strong. The numbers are counterintuitive, however; as the number gets bigger, the wire gets thinner; thus, chicken wire is quite thin. As you might now guess, a 14-gauge wire is thicker than chicken wire, and 12-gauge wire is thicker still.

Why do smaller numbers indicate thicker wire? With early measuring systems, the number indicated how many times the original piece of wire (0-gauge) had been heated and pulled through a hole to slim it down. A 14-gauge wire had been slimmed down 14 times, and a 20-gauge wire had been slimmed down 20 times.

Galvanized. Fencing is made of steel wires, but when exposed to moisture, steel wire alone simply rusts and crumbles away in a few years. Steel fencing wire is commonly galvanized, or coated with a grayish sliver layer of zinc, to slow down rust. It's not perfect, however: you don't want to bury fence wire even when it's galvanized. Over time, in damp conditions, organic acids in the soil will dissolve the zinc. When that protective layer is gone, the steel is exposed to air and water and rusts away. This point will be important when we talk about antipredator aprons.

Like galvanized fence wire, galvanized feed cans and water troughs will also eventually rust away where they touch the soil. Stand them on bricks, cinder blocks, or slabs of rot-resistant wood to get them off the ground.

Welded wire. In these fences, the horizontal wires are spot-welded to the vertical wires. Fences of the same gauge that are woven or knotted instead of welded are stronger, but more expensive. A welded wire fence is plenty strong enough if the gauge is at least 14. Don't let your heart get broken; mount a good defense by knowing which is the best fence for your chickens.

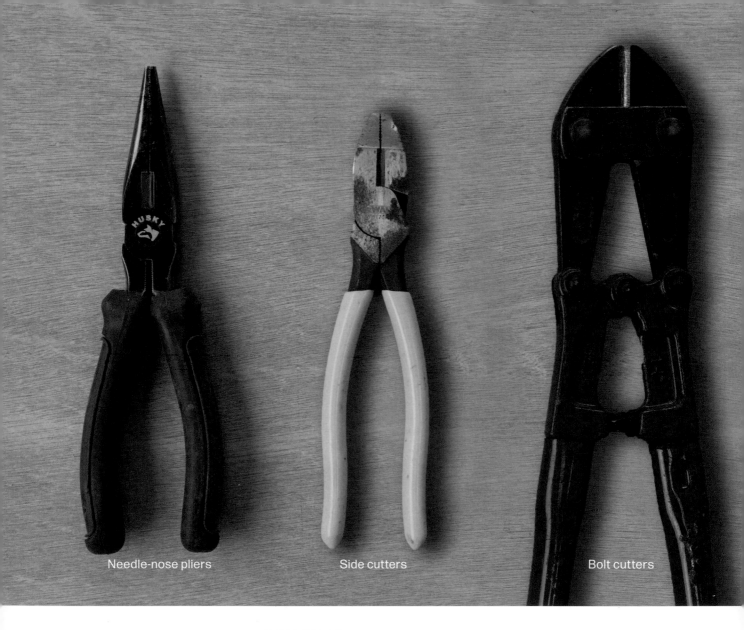

Needle-nose pliers

Side cutters

Bolt cutters

CUTTERS

There are a variety of hand tools for cutting fencing to size: bolt cutters, side cutters, needle-nose pliers. Even most garden pruners have a spot at the base of the cutting blade that's shaped for cutting wires as thin as chicken wire. It's worth remembering that the heavier the tool, the quicker the cutting, and the less wear and tear on your hands.

If you want to speed up the cutting process, an inexpensive metal-cutting blade for circular saws will allow quick work. It has a smooth cutting edge rather than teeth so it won't snag on the wires. Metal-cutting blades for jigsaws and reciprocating saws have teeth; they'll work, but things can be a little herky-jerky when cutting fence wire.

Needle-nose pliers are handy for cutting 14-gauge wire fencing. They are also helpful in holding fence staples as they are hammered in.

The heavier the tool, the quicker the cutting, and the less wear and tear on your hands.

FENCE STAPLES

In most of the projects in this book the preferred type of fastener, especially for novices, will be a screw rather than a nail. For fencing, a better fastener is a ¾" fence staple. As with fencing, staples will last longer if they are galvanized. Most staples are U- or V-shaped, with pointed ends. To secure a section of fence wire to a wooden post, position the points of the fence staple so that they straddle the wire, then hammer the staple in place. If you find it challenging to drive a nail, hold the staple in position with needle-nose pliers as you whack it with the hammer.

Alternative: Nails will work in a pinch if they are at least 1½ inches long. If you have a bunch of nails handy and want to save the expense of buying staples, drive a nail next to a strand of fence wire, but only about halfway in. Then use the hammer to bend the nail over the strand and then flat against the post so that it holds the fence wire tightly in place. It's not pretty, but it's highly functional and cost-effective.

A galvanized ¾" fence staple is the best way to attach fencing to a post or the coop.

HOW TO MAKE IT

FENCING THE PEN

Attaching fencing can be a one-person job, but it's easier with two: one to hold the fencing upright and snug against the posts, and one to drive the staples. I recommend cutting it into sections long enough to reach from one post to the next.

WHAT YOU NEED

- Roll of 14-gauge galvanized welded-wire fencing, galvanized or coated with black plastic, of a length and height to fit your pen

- Measuring tape

- Cinderblocks or bricks

- Needle-nose pliers, circular saw with metal-cutting blade, or other cutting tool

- Galvanized ¾" fence staples, U-shaped or V-shaped

- Hammer

WHAT TO DO

1. Cut the fencing to fit.
Roll the wire out on the ground, using bricks or cinder blocks to keep it from rolling back up. Measure the distances you'll need for each section of wire before you cut, making sure each section will overlap the face of the post at each end.

2. Attach the top of the fencing to the first two posts.
Staple the top of the piece of fencing to the tops of the first and second post so it's snug and straight. You can use pliers to hold the staples while you start them. The bottom of the fence should just meet the soil surface.

3. Add staples down the length of both posts.

Set a staple at about every handspan (the distance from the tip of your thumb to the tip of your little finger with your hand spread as wide as possible, or about 7 to 8 inches).

4. Cut more fencing and repeat for each pair of posts.

Keep the fence as snug and straight as possible so that it doesn't sag, which not only looks sloppy but also makes it hard to drive staples as you go down the line.

Keeping Diggers Out

Some people recommend installing fencing below ground to keep predators from digging a tunnel under the fence. In the soil, however, as noted on page 55, even galvanized fencing will last only 5 to 10 years (depending on moisture levels) before it rusts away. Since you can't see that, you won't know if your run has become vulnerable to digging predators.

A better defense is to lay an apron of fence wire on the ground at the base of the fence that surrounds your pen. When predators approach the fence, they will try to dig right next to it but will be foiled by the apron. Yes, the apron also will eventually rust away, but it will be easier to replace than fencing buried under the ground.

If you can scavenge some fencing, your apron won't cost you anything. Metal is never trash. Over many years, I have collected miles of leftover fence wire from the curbside. Most I reuse, sell at the scrapyard, or pass on to a friend for a project. At home, I use a variety of scavenged fence wire to keep critters from digging under the fence into our chicken run.

A better defense is to lay an apron of fence wire on the ground at the base of the fence that surrounds your pen.

ANTIPREDATOR APRON

WHAT YOU NEED

- 2-foot-tall section of fencing at the length your pen requires

- Flat board

- Metal snips, bolt cutter, or side cutter

- Mulch

CONSTRUCTION TIP

If the fencing material you have on hand is 4 to 6 feet tall, you can cut it down to 2 feet tall with any metal-cutting tool.

WHAT TO DO

Cut the fencing and bend it into an "L" shape.

1. Roll out a 2-foot-tall roll of fencing and cut it into easy-to-handle 2- to 4-foot lengths. These sections will overlap a bit as you work your way around the pen.

2. Bend each short section into an L shape, with "legs" of about 8" and 16". It may be easier to bend the fencing along the edge of a board.

Give your defense some "teeth."

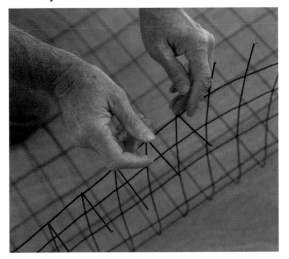

3. Cut away the lengthwise wires at the top and bottom of the apron to leave prongs. Bend the prongs on both edges out at a right angle (90°).

Wrap the apron around the base of the pen.

4. Set the fencing so that the 8" section of the L is attached vertically to the outside of the pen fence. The longer 16" side of the L lies flat on the ground. Once in place the L will look as if it's lying on its back, with the short side pointing up and resting flat against the run fence.

Secure the top of the apron to the run fence.

5. Bend the top prongs over the pen fence wires.

Anchor the bottom of the apron to the ground.

6. Wet the ground and press the bottom prongs into the moist soil. (Dry soil may be hard to penetrate.) Cover with mulch.

Other Types of Fencing

Depending on your chickens' needs and your wishes, there are other options for safely containing your flock.

Hardware Cloth

Some henkeepers with a larger budget and extra time use hardware cloth to fence their run. This is not, in fact, cloth at all, but fencing material of very sturdy steel with a coating of zinc — in other words, a galvanized fence of welded wire. The small openings are ¼" × ¼" or ½" × ½" squares.

When working with this material, wear leather work gloves: hardware cloth is stiff yet springy, and it likes to bite. It is also about three times more expensive than 14-gauge wire for the same coverage.

On the plus side, it is a good material to cover openings in your coop. It allows good ventilation (important in all four seasons), yet the openings are small enough to keep snakes, mice, rats, and predatory birds out of the coop and away from the eggs.

Heavy, expensive hardware cloth, like this metal mesh, is best used to cover windows and gables in the coop. Its openings are small enough to block snakes, mice, rats, and predatory birds.

Chain-Link Fencing

Chain-link fencing is familiar, with its diamond-shaped openings. It's commonly used for property-line fences, but because it is a much heavier gauge than welded-wire fencing, it costs a lot more. Unless you have access to some free chain-link fencing, such as that from an old dog kennel, I wouldn't recommend using it for your run fence. Welded wire costs less and will be strong enough.

Create a Chicken Corral

You might be tempted to let your chickens forage in your yard so that they can eat bugs and weeds. Yes, some bad insects and seeds will be devoured, but at what cost? Well, just these eight little problems, which will plague small gardens in a very short time:

1. The lawn will be gone.

2. Mulched beds will be scratched bare.

3. Shallow bulbs will be uprooted.

4. Annual flowers will be massacred.

5. Perennial flowers will get beheaded.

6. Patios will be pooped on.

7. Decks will be pooped on.

8. Chairs and benches will be "fowled" up.

A brace of chickens would be thrilled with the resulting swept-dirt yard punctuated by a few hardy shrubs. But most gardeners wouldn't be too happy.

Domenica, Mezza Luna, and Stella d'Oro believe the weeds are always greener on the other side of the corral fence.

Short of building fences around every garden bed or patio, what's a gardener and pro-foraging chicken-keeper to do? The answer: Install a simple chicken corral where your hens can forage without destroying your gardens. Plantings of established shrubs and trees are big enough that they can withstand foraging. Their branches will protect your hens from hawks, too. If you can corral your chickens there, the rest of your yard can thrive.

After we got fed up with the amount of destruction our three hens wreaked free-ranging in the garden (which took just a few days), we decided to sequester them in a bed composed of five mature ornamental shrubs. With a location selected, I needed a way to corral the birds. I wanted a secure enclosure that either looked good or was nearly invisible. But I also wanted to save time and money. Sounds like a set of conflicting goals, but that sort of thing often inspires the best results.

Since a 4-foot chain-link fence already encloses our yard, I didn't need something strong enough to keep stray dogs out. I just needed something stable enough to keep the hens in during the day. A conventional fence with a gate and posts would have looked nice but would have been expensive and time-consuming to build. A prefabricated electric fence wouldn't have been cheap either, and the white strands of wire would have been an eyesore.

I settled on using a couple of 3' × 50' rolls of 14-gauge welded wire fencing that's coated with black plastic. This fencing has 2" × 4" openings, the same as recommended for the hen pen. It's long-lasting but cheap, costing less than $1 per linear foot. And since black objects reflect little light, the fence doesn't catch your eye from a distance.

In the open, chickens can fly over a fence only 3 feet high. By running the fencing along the outer branches of the shrubs, I created a "no-fly zone" along the top of the fencing. The shrubs' branches also keep hawks and other predatory birds at bay. At that height, I could just step over the fence on the rare occasions I needed to get inside. Using a few designer tricks, I installed this fence in a couple of hours without posts or a gate. We call our corral the Forage Grove. The chickens are happy, and the rest of our yard has recovered.

By combining the Forage Grove with the Chunnel (page 68), and investing just a small amount of money and time, you can sit in a garden chair and enjoy watching your hens scratch around in their very own corralled paradise.

Here's how to install this low-tech, low-cost, easy-build enclosure.

I wanted a secure enclosure that either looked good or was nearly invisible. But I also wanted to save time and money.

HOW TO MAKE IT

CHICKEN CORRAL, AKA FORAGE GROVE

WHAT YOU NEED

- Enough rolls (a standard roll is 3' × 50') of 14-gauge welded-wire black plastic–coated fencing to enclose a shrubby area

- Circular saw with metal-cutting blade

- Pliers with wire-cutting blades

WHAT TO DO

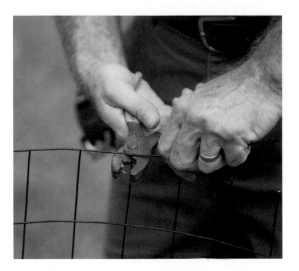

1. Before unrolling the fencing, remove the lowest horizontal strand of wire.

Use the circular saw with the metal-cutting blade, or the pliers, to create prongs at the bottom edge of the fence. After step 4, these prongs will be pressed straight into moist soil; that will anchor the fence and make it 4 inches shorter, thus easier to step over.

2. Roll out the fencing and cut it into sections.

Using the pliers, cut the fence into more manageable sections, about 4 feet long. That length also makes it easier for the fence to follow the slope of the ground as it rises or falls.

Note: The wire is springy, so you may need to set something heavy on it to keep it from rolling back up.

3. Prepare the sections.

Use the pliers to cut off the vertical strand of wire at the right-hand end of each section, leaving prongs. To assemble the fence, you'll bend these prongs around the vertical wire at the left side of the next section, as shown here.

4. Cut foot spaces at the bottom of the fence.

Every 12 to 18 inches, snip a couple of vertical strands for your feet between the two horizontal wires just above the prongs, and bend these strands up and out of the way.

5. Set up the fence.

Hold each fence piece in place and step into the openings you've cut to push the bottom prongs into the ground. Bend the side prongs to tie the fence pieces together as you go.

CONSTRUCTION TIP

If you have a big dog who thinks chicken poop should be on the menu (I'm looking at you, Abbey), then parts of this simple fence may get trampled or uprooted when you're not looking. If so, use 2' lengths of rebar (inexpensive steel rods available at building supply stores) and weave them vertically through an opening in the fence. Use a hammer to drive a few stakes into the ground as needed to stiffen the corral fence.

Poppy enters the Chunnel through an opening at the base of the pen. This opening gets closed at night or when we go on vacation.

A Chunnel for the Lunchtime Commute

With the corral completed, it is now a matter of getting the hens from their hen pen to this idyllic Forage Grove. You don't want to carry the chickens from pen to corral by way of the gate. You want them to carry themselves: more exercise for them, less hassle for you. In my case, the chicken corral was about 25 feet away from the pen's gate, but the corral was less than 6 feet away from the closest part of the pen. Hmmm.

The solution is to install a small, chicken-sized gate, with hinges and a latch, at the point in the pen nearest the corral. Then cut a chicken-sized opening into the corral fence.

Linking the hen pen to the corral with a "Chunnel" — short for chicken tunnel — allows the birds to come and go on their own. You can move the Chunnel aside temporarily to get a wheelbarrow by. When you don't want the chickens in the corral, simply move the Chunnel and close the small gate to the hen pen.

HOW TO MAKE IT

CHUNNEL

WHAT YOU NEED

- Measuring tape

- Pliers with wire-cutting blades

- 3-foot-tall 14-gauge welded-wire black-vinyl-coated fencing (the same fencing material used for the Forage Grove)

WHAT TO DO

1. Using your tape measure and pliers, cut a section of fence wire long enough for your chunnel.

2. Snip off the strands of wire on each side of the Chunnel that will touch the ground. Cut some slots for your feet, as in step 4 on page 67. Then snip off the strand of fencing at each end of the Chunnel where it will meet other fencing or sections of Chunnel.

3. Shape the section of fence into a tunnel resembling the photo on page 68.

4. Bend the prongs on the end(s) of the Chunnel to attach them to corral fencing (or other sections of Chunnel).

5. The end of the Chunnel that meets a small gate in the pen should be positioned close enough to channel the chickens, but loose enough that you can move it to shut the gate at night or when on vacation.

6. Use pliers to snip out some wire in the corral fencing so that your hens can pass from the Chunnel into the corral.

Avian Alcatraz

A client wanted to keep her hens from escaping. They were jumping to the top of her 42" picket fence that enclosed their run. I could have built a much taller picket fence, but it would have cost a lot and the pen would have looked like a fortress. My helper Dave Mauldin did some online research and came up with a solution that he calls "Avian Alcatraz" for its resemblance to the angled arms and barbed wire at the top of prison walls.

A less punitive description would be to say it looks like a clothesline running along the inside of the fence. Wooden arms near the top of each post reach inward horizontally about 18 inches. Two strands of heavy-duty weed trimmer line are passed loosely between each arm.

Dave didn't want to use fishing line or twine, as he saw in some of the online plans he found. A wing caught up in some fishing line can come away cut and bloody, and twine will break down from the rain and sun. Dave's choice — heavy-duty weed trimmer line, preferably of a dull gray rather than a loud orange or green — won't cut flesh and should survive the elements for a couple of years before needing replacement.

It worked right away. The hens want to stand at the base of the fence and flap their way to a landing on top. But the trimmer lines block their path, and the lines aren't taut enough for the hens to comfortably land on either. These "yardbirds" may not have committed any crime, but they do stay in their pen now.

A 42"-high picket fence will keep hens (even guinea hens) in if bolstered by Avian Alcatraz fencing. Use heavy-duty weed trimmer line, strung loosely.

4 Enrich and Accessorize

When I was a young man I worked at an egg farm. For one day. The environment was all bleakness interrupted by concrete. Not fit for man nor beast. Our Hentopia looks nothing like that egg farm. While maybe our chickens don't recognize the many little touches we add to their domain, we and our guests sure do. I think the hens do appreciate having a tree to perch in and crickets to chase. So we keep adding little improvements for them and us. We're always happy to have one more reason to cruise some junk shops.

Playtime

Zookeepers often point out the "enrichment" features enlivening each animal's enclosure at the zoo, such as ropes for monkeys to climb or indestructible balls for pandas and bears. The idea is to let the animals entertain themselves while they wait for dinner (which can never come soon enough).

Chickens like to have a little fun, too. For birds that don't get to spend every waking moment free-ranging, it helps their general happiness and productivity when we provide a few enrichment features in their pens.

To me, part of the fun of any new adventure comes from acquiring the proper accessories: items that enrich the experience for us as well as for our chickens. That might be functional things like funky feeders or fancy gates, as well as fun items bought at junk shops or boutiques, like vintage signs, antique drawer pulls, Tibetan prayer flags, or farm animal sculptures — whatever strikes your fancy. Here are some tips on creating a playful hen habitat.

Signs, flowers, sculptures, murals, and tchotchkes from any or all nations are suitable accessories for a custom coop.

Scratch-and-Snack Bar

Chickens like to scratch the ground looking for bugs and seeds to eat. It's what they do instead of heading into the office or catching up on social media, and it's called foraging. Most of our yard is off-limits for foraging, as chickens can destroy a lawn or flower bed quicker than you can mistakenly hit "reply all" on a snarky office e-mail.

Most of the time our hens forage in the pen, but since the pen is a small space we enrich it a couple of times a year with free wood chips dumped in our driveway (on request!) by local tree trimmers. Adding a 4- to 8" layer of wood chips to the pen provides a food source for soil insects that happily sacrifice themselves for our chickens' welfare. (At least I think the bugs are happy. The chickens certainly are. Every bug they eat is that much less feed I have to buy, which makes me happy, too.) The decomposing wood chips also provide an environment for microbes that quickly decompose the chicken poop in the pen.

We also throw leftover garden seeds into the mulch, and our girls come running. Scratching and searching for treats in the wood chip snack bar is one of the high points of their day. As the hens scratch, the seeds get mixed up in the mulch, which creates a more challenging process for the hens and makes it last longer than tossing seeds onto bare dirt.

Scratching and searching for treats in the wood chip snack bar is one of the high points of their day.

Henkeepers realize another benefit from the wood chips: the pen isn't muddy when we need to top up a feeder or do any other chores. Some readers may be thinking, "Our pen is never muddy. We have a roof over it." A roofed pen is not uncommon, and in areas with long snow seasons it may be necessary. Chickens don't enjoy the snow and, depending on the depth, rarely scratch their way down to the soil.

The downside to a roofed pen is that the ground underneath will be dry as a desert, too dry for the poop to decompose, and unfit for any bugs that want to sacrifice themselves as a crunchy afternoon snack. The chickens will still scratch the dry soil, but they'll be as disappointed as prospectors at a played-out gold mine. Even in snowy regions, it may be worthwhile to leave part of the pen with just fence wire as a roof, so that rain can wet the ground during the rest of the year.

Yet another benefit henkeepers can realize from enriching the pen with wood chips is compost. After four years, the ground level of our pen had risen about 8 inches with a rich layer of composted chicken poop and wood chips. I shoveled that layer into a wheelbarrow and spread it on our vegetable garden. It took me only a couple of hours, and I could hear the garden's sighs of delight and whispered promises of future bumper crops.

Fungi and the Fowl

Edible mushrooms (called wine caps — *Stropharia rugoso-annulata*) grow in our wood chip mulch, and we make full use of them. They get their name both from their wine-red tops (which turn silvery gray in the sun) and their flavor of potatoes cooked in wine. Like a comic-book hero, wine caps also possess the superpower of being able to destroy *E. coli* bacteria. Fortunately, they are one of the easiest mushrooms for a novice to grow.

Chickens don't eat mushrooms, and the wine caps don't contain any *E. coli*. Nonetheless, all mushrooms should be cooked before being eaten. Growing wine caps isn't rocket science, but the steps involved are beyond the scope of this book. Mycologist and chicken-keeper Tradd Cotter and his wife, Olga, introduced me to their method. They run a business called Mushroom Mountain, and on their website (see References and Resources) you can order the mycelia for wine caps and see Tradd's detailed drawings and instructions. What I can share is my recommendation for the best way to eat wine caps: washed, sliced, seasoned, briefly sautéed in butter, and then cooked in an omelet of fresh eggs from your own happy hens.

Stella d'Oro watches the world from the tree trunk.

Jungle Gym

As any henkeeper quickly learns, even an upside-down pie pan will entice a chicken to hop onto it to get a better view of her surroundings or to assert her dominance in the hen hierarchy. They do it because they're channeling their ancestral instincts.

Chickens evolved as forest birds in Southeast Asia. Sometimes during the day, they'd hop up into the lower branches of trees to scout for dangers, or just to feel like they were the lords of all they surveyed. Later on, after a good dinner of crunchy bugs, rich seeds, and succulent greens, they'd hop back up into a tree to roost the night away, safe from predators.

To support that daytime behavior, I installed the twisting trunk of a dead juniper tree at a quirky angle. I clipped away the smaller twigs and left the sturdy skeleton of the trunk and some side branches thick enough to support a well-fed bird. I anchored the bottom of the trunk in the soil and lodged the top against a rafter that ran across the roof of the pen.

Three of the branches show erosion of the bark that indicates regular use. We enjoy seeing Buttercup or Poppy up high, taking a good look around, or hunkered down for a rest. Being juniper, the trunk is rot-resistant, but any species of tree will last a few years and give your girls those long views they seek.

Leafy Office Memos for the Girls

A clump of pigweed in the binder clip creates a feeding frenzy.

Years ago I remember one of our local chicken gurus, Bob Davis, saying that he had started clamping garden greens inside the pen fence to make it easier for his girls to eat. I didn't think much of it at the time, but I remembered his comments when I saw a large office binder clip on Chris's desk.

Certainly, chickens can eat small items, like pieces of grain, off the ground. But larger things like cabbage leaves or bunches of carrot greens can be trickier. Chickens evolved tugging leaves off rooted plants; if the leaves don't fight back, chickens will have a harder time breaking off a piece small enough to eat. They will probably never figure out that standing on the leaves will make it easier to tug away with success. Whenever I simply tossed greens on the ground I got used to seeing lots of them going uneaten.

Then I saw the big office binder clip on Chris's desk, and I quickly envisioned a contraption like Bob's. I scooped up the binder clip and scouted the garage for a 6" length of tie wire (twine, zip ties, or anything similar will do). I loaded the clip with a bunch of overmature kale

Chickens evolved tugging leaves off rooted plants.

and tied the clip's handle to the fencing of the gate at head-height for a chicken.

I positioned the binder so that we could crouch down, open the gate, and with one hand outside the gate pinch the binder open. With the other hand inside the gate, we could slip that day's leafy memo into the mouth of the binder clip — all while keeping the girls from making a great escape, of course.

With the gate closed, I watched the tasty tug-of-war begin. Our hens really went to work, going after the beetles on the kale (the garden's organic). As soon as they had hoovered up the beetles, they quickly ripped the kale down to its skeletal stalks.

You're welcome, girls.

Buttercup studies the presentation of today's hors-d'oeuvres.

Playpen / Isolation Ward

Violetta sojourns in a dog crate to break her from a broody spell. The lid from a plastic tote gives her midday shade.

The coop and the hen pen are two enclosures necessary for proper chicken-keeping. In addition, I highly recommend acquiring a third one: a dandy device known as a dog crate. This accessory falls into the category of functional rather than fun: you'll want it for isolating one or more birds.

There are at least four occasions when a henkeeper will have need of an isolation chamber:

1. Periodically, hens get broody and need to be kept away from the nest box to break their mood.

2. Also, any adopted birds will need a safe space of their own in or next to the pen before they're accepted as members of the flock. Your original birds will peck any birds they see as "intruders" if they can get to them. But with a secure home close by, your original birds can get acclimated to the sight, sound, and smell of the newcomers.

3. In a worst-case scenario, a hen may have an infectious disease and need to be quarantined. Cast-off dog crates are perfect for that purpose.

4. When baby chicks are big enough to move into the pen a crate makes a safe halfway house for them until the older hens are ready to accept them.

Many years ago, when I saw a discarded yet functional metal dog crate by the side of the road, I suspected it would come in handy for something at some point. Into the back of the garage it went. A year later we adopted a mature Black Marans hen and named her Dahlia. We needed her to get acquainted with the old girls without being pecked like a piñata. Bring out the dog crate.

A new or salvaged dog crate is a safe way to isolate birds. Be sure to provide a waterer and feeder as well as cover from sun and rain.

The crate was big enough to hold a Labrador retriever, so it had plenty of room for a single hen. The heavy-duty wire cage had a door in one end and a hinged top panel too (cleverly designed to lie flat for storage.)

We wanted our new hen to be able to live in the crate for a week or two, until our original birds got used to her. We put the crate inside the pen with a couple of bricks at one end to support a store-bought waterer and feeder at chest height for Dahlia. That put them high enough that any scratched debris wouldn't clutter them up. A tarp over the top kept out rain and direct sun.

It did feel a bit like punishment to sequester a hen in a dog crate even though Dahlia had room to move around and still be within sight and smell of her new flock. She also had plenty of shelter, food, and water.

The original birds, Buttercup and Poppy, weren't happy about the newcomer, but they couldn't get at her. After about 10 days, they stopped paying attention to Dahlia and we let her out to join the flock with no bloodshed. Mission accomplished.

We still have the crate. Between active uses, it functions as a bin that holds the waterer, feeder, roof tarp, and other paraphernalia, including a salvaged cat carrier that we used to transport Dahlia in the car.

A year later, Dahlia got broody in the nest box — she mistakenly thought she could hatch some eggs. With no roosters, the eggs are infertile and therefore won't hatch; and if the eggs don't hatch, a broody hen never gets the message to leave the nest. Under those conditions she could suffer or even die from dehydration or starvation. A henkeeper must break this mood to save the hen.

We were ready. We popped Dahlia into the dog crate for her own good. After a week, we let her out, but she went right back to brooding in the nest box. We hadn't waited long enough. Another trip to the dog crate for another week did the trick.

As our old girls have aged and drop fewer eggs, we've raised four new chicks to adulthood: Domenica, Mezza Luna, Stella d'Oro, and Violetta. When they were chicks, the dog crate gave them a safe sanctuary in which to be near our older hens until they were recognized as part of the flock.

Dahlia had room to move around and still be within sight and smell of her new flock.

Cricket Rodeo

As the top hen, Buttercup gets first crack at the cricket crack while Mezza Luna waits her turn. Rest assured, chickens are not vegetarians (or egalitarians or pacifists either).

It's been decades since I've done any fishing, but I still keep a cricket bucket rolling around behind the seat of my pickup truck. The fish have nothing to fear from me when I buy a few dollars' worth of crickets at the bait shop, but I know the chickens will be very excited. They enjoy any worms or grasshoppers we toss their way, but they sure love them crickets.

A cricket bucket has a solid plastic bottom, but the sides are made of screen to allow the crickets to breathe. The cleverly designed top has a flange that lets you reach a hand in to catch a cricket, but that keeps the crickets from being able to climb out. I picked it up for fifty cents at a yard sale. I probably could have asked to buy it for a quarter, but the sellers looked to be having a slow Saturday.

I reach into the bucket and toss a dozen crickets into the pen to get the chickens running around. They're stretching their necks and snagging crickets off the ground and off each other's backs. It's like watching cowboys catching calves at a rodeo. Then I put the cricket bucket down on its side and let the girls at it. Being unable to get at them through the screen confuses the hens at first, but they eventually see how this game works. Buttercup is dominant, so she has the first go, but the others get their licks in, poking their heads into the bucket and coming out with juicy, crunchy prizes.

Of course, if you don't live near any bait shops, you can still buy crickets at any pet store that supplies them for feeding pet lizards. Chickens did evolve from dinosaurs, which is Latin for "terrible lizards." Therefore the crickets are fair game for a quick-and-dirty lizard rodeo.

A fishing accessory like this cricket bucket makes a good corral for the cricket rodeo.

HENKEEPERS I'VE KNOWN

Adventurer Hans Voerman's Caribbean Hen Habitat

The small town of Rincon on the tropical island of Bonaire

When Hans Voerman was a child in Holland in the 1970s, stories about deforestation and dying coral reefs made him angry and sad. He had fantasies about protecting wild lands from destruction. As a young man, though, he couldn't find opportunities in that line of work.

Hans served as a maritime survival instructor in the Dutch Navy and as a civilian as second mate on a tramp freighter, helping with docking and navigation while the ship hauled salt, timber, tobacco, pumpkins, onions, used cars, and once an entire factory to ports around the globe. In his travels he saw ruined lands as well as lush areas, and the experience reinforced his desire to find a piece of ground that he could preserve, care for, and make his living from.

Today he protects a 1.5-acre property in the backcountry of the Caribbean island of Bonaire. He plants native trees that support Amazonian parrots, Caribbean parakeets, orioles, emerald hummingbirds, bats, and other hardy wildlife on his desert island. To be self-sufficient, he also runs a small eco-lodge, leads outdoor tours, and keeps ducks, pigs, and chickens.

Finding Bonaire

Hans discovered Bonaire in the early '90s on a visit to his former scuba team officer. They had served together in the Dutch military, where Hans learned to handle underwater explosives. The weeklong visit turned into a month, and a month turned into a life choice as Hans settled on Bonaire.

Like many Caribbean islands, Bonaire suffers from the overgrazing caused by free-ranging goats and feral donkeys, and its growing popularity has led to development pressure that concerns local environmentalists. "Every day we lose trees on this island," Hans says. He stays active in environmental debates and educates locals and tourists at his lodge and on his tours.

Enter the Chickens

Before coming to Bonaire, Hans had no experience with animals: "None, absolutely zero. But I always liked to be self-sufficient. It's relaxing to watch and take care of animals. And to occasionally eat them." About 10 years ago a friend gave him some chicks to start a flock. He keeps a rooster and often has 20 to 25 birds of all ages, of a hardy local breed that everyone just calls a Bonairean chicken.

The rooster has some of the coloration of a Dutch Bantam, but without a doubt other breeds are mixed in. Dutch Bantams are from Indonesia, a former Dutch colony. It's said they are as close as you can get to the look of the original red jungle fowl, but Hans's birds are bigger than bantams and lay a normal-sized egg. They probably have an interesting heritage, given that the Spanish, the Dutch, and the U.S. Army (during World War II) have all governed the island.

Hans doesn't worry about critters like raccoons, foxes, or possums, because there aren't any carnivorous, four-legged mammals on Bonaire . . . except for stray dogs. When he bought the land, he installed waist-high wire fencing on the property line. That keeps out the dogs, along with the goats and donkeys. With the grazing animals at bay, Hans's land resembles a lush, square oasis in a nibbled-down desert of small trees, candle cactus, agaves, and scruffy grass.

For extra protection Hans built a cinder block coop for the birds to roost in and get out of the occasional rains. One side of the coop opens into a pen lined with window screen and wooden lath. The screen and lath are strong enough to keep out the main intruders: omnivorous, cat-sized, blue-tailed lizards that would eat the kitchen scraps, grain, and perhaps the eggs.

The screen and lath are strong enough to keep out the main intruder: omnivorous, cat-sized, blue-tailed lizards.

These aren't the only threats to chickens. Another island farmer lost all his birds a few years ago when he tried to eliminate a wild honey bee colony in a storage locker. The riled-up bees, probably Africanized, attacked, stung, and killed all the chickens. The farmer retreated to his car to nurse just a few stings, but has a new flock now and no wild bees on the grounds.

Tropical Zone

In Bonaire's dry weather with a nearly year-round high of about 81°F (27°C), the chickens rarely have disease or pest problems. Infrequently, Hans says, one bird will seem a little strange — slow or quiet — and then will die the next day. So they don't burden him with vet bills.

Since the island is only 12 degrees north of the equator, day length doesn't change enough to affect egg production. The birds are hardy, but for Hans, they aren't high producers. He reports getting "two to three" eggs a day from his flock, "sometimes more, sometimes nothing. If you have eggs for a while you have to save them up, as sometimes they don't produce anything."

He periodically slaughters an older bird for meat. Because they are older and tough, he cooks them in a pressure cooker with half a kiwifruit, which helps tenderize the meat. These birds mostly go into a soup, or he dices the meat into fine cubes to go with rice or pasta.

Occasionally his guests at the eco-lodge take an interest in the flock and help with the feeding and watering. A former girlfriend had a deathly fear of chickens. Because she was the one who fed them whenever he was away from the lodge, Hans set up a special device so she wouldn't have to enter the pen. He poked a 6-inch-diameter, 8-foot-long PVC pipe at a jaunty 45-degree angle through the wall of the chicken pen, creating, in effect, a dumbwaiter or laundry chute for these classy birds. His girlfriend could toss the feed and kitchen scraps into the pipe, and they would slide down to land on the dirt floor of the pen. A tight-fitting lid on the end of the pipe kept rain and critters out, and she could feed the chickens without having to enter the pen or even go near them.

A Piece of Paradise

After decades of relying on off-island production, Bonaire is seeing agriculture slowly return. Some back-to-the-landers are growing sorghum in the dry climate, for example, but there aren't enough island-grown grains for Hans's chickens. He could buy layer feed from Venezuela

> **In Bonaire's dry weather with a nearly year-round high of about 81°F (27°C), the chickens rarely have disease or pest problems.**

or the Dominican Republic. To feed his animals grain that's free of antibiotics, however, he buys feed imported from Holland at $19 for a 55-pound bag, supplementing it with kitchen scraps.

On his land, he allows a small tree called French cotton (*Calotropis procera*) to grow, and after a rain he harvests the leaves to feed the animals. Not a true cotton plant, it's a milkweed from Africa with a silky fluff, and a favorite of the monarch butterfly.

Hans built the lodge, where he lives and keeps his animals, with his own hands. It has a stone foundation, rustic ceiling beams, and a breezy second-floor porch with a dining area, a hammock, and a hanging chair.

The eco-lodge uses solar panels for electricity, a windmill to pump well water, and roof-mounted barrels to heat the water and create water pressure for the kitchen and the indoor and outdoor showers. The roof also supports a stargazing platform, from which you can also see owls, bats, and nocturnal birds called nightjars. This distant from distracting urban lights, one can also see meteors, the Milky Way, and the constellation Auriga, the chariot driver.

Hans chose Auriga as the name of his eco-lodge because his last name, Voerman, also means chariot driver. When not tending his lodge or his animals, Hans drives a modern chariot in the form of a black, four-wheel-drive, four-door pickup truck with a bright red bat painted on the fender.

When Hans bought his land a dozen years ago, it was a barren, goat-eaten bit of scrub. By pursuing his dream of protecting nature, Hans, like the Dutch boy with his finger in the dike, is helping to hold back the flood of development. And he's making a cozy home for himself, his guests, and his chickens at the same time.

For more information on Hans, his eco-lodge, and his tour business, see the References and Resources section.

Between caving tours, Hans Voerman inspects one of his Bonairean chickens.

PART II
INS AND OUTS OF CHICKEN COOPS

The word "coop" comes from the Old English word for "basket." To a large degree a chicken coop is simply a basket that your hens sleep in. But lots of things have to be able to get in and out of it: chickens, eggs, the henkeepers, fresh bedding, spent bedding, hot air, damp air. Let's look at how we'll make all that possible.

5 Doorways Four Ways

Even for a small coop and hen habitat you'll want at least four doors: (1) a gate into the pen; (2) a chicken-sized automatic or pop door from the pen into the coop; (3) a keeper-access or henkeeper's door, also from the pen into the coop; and (4) at least one if not two big coop-cleanout doors on the perimeter of the pen for scooping poop into compost bins.

Which Way Should They Open?

For the doors on the coop, it's better that they open *out*. If they were to open *in*, they would encounter the bedding and poop. Plus you'd have to leave space at the bottom so that they could swing in above the floor, and that would create drafts and a gap for snakes and mice to enter.

The same issues apply to a gate: an in-swinging gate will be hard to open if the hens have scratched a bunch of debris in front of it. Installing gates and doors to open *out* also makes it easier to install their hinges.

Don't Come Unhinged

Hinges will be exposed to the weather, so buy *exterior*-grade hinges that won't rust and bind up. Save some time and hassle: buy hinges that come with a matching set of screws.

In most cases, two hinges will be enough for any door; placing one near the top and one near the bottom will help keep a door from warping out of shape. On a tall gate it doesn't hurt to add a third hinge at the midpoint.

To every door a purpose. This door Is for chicken-keepers to access the interior of the coop for maintenance and adjustments.

This kind of latch is called a swivel safety hasp and would be hard for a raccoon to open. These hasps are designed to be used with a padlock, but they're complicated enough to keep raccoons out by themselves.

Latch Me If You Can

There are lots of options for the latches that hold the doors in your pen shut. Hardware stores and big-box home-improvement stores will offer a couple of inexpensive choices. One is the hook-and-eye latch that's common on screen doors. It's simple enough for raccoons to open, unless you get the kind with a spring-loaded cover. Another is what's labeled a "gate latch." These require two actions to open, so they're too complicated for raccoons. Avoid sliding-bolt latches: they require that the door remain in a precise position, guaranteeing future frustration.

I recommend store-bought latches for gates, but for the keeper-access door inside the pen, simple handmade latches of scrap wood are fine if you want to save some money (more about those below). For access doors on the perimeter of the pen, such as the cleanout doors, where raccoons can stand on the compost bins and reach the latches, I recommend what's called a "swivel safety hasp," as shown in the photo. You have to swivel the oval part and then swing the hasp open. That's easy for a person, but too complicated for a raccoon.

Handle It

Handles aren't absolutely necessary. You can forgo them, buy simple or fancy ones from a hardware store or thrift shop, or use a curving piece of a cedar branch that's the right size and shape for a rustic handle. I have a handle on my gate but not on my coop doors. My plan has been to make some handles with that clever rustic look, but the task hasn't climbed its way onto the top of my to-do list just yet.

If you do want to make your own handles from cedar branches, just drill two holes, one near the top and one near the bottom. Make the holes big enough for a screw to slide through easily. Mark the position of the holes on the door with a pencil or the tip of the screw. Predrill holes the same diameter as the shaft so that the threads will bite the wood. Install and enjoy.

Great Gates for Great Girls

Part of the reason any of us keep chickens is that poultry have personality. Our chickens' pens and coops should have a little personality, too. As far as pens go, the best place for adding a bit of character is the gate — but gates are one of the trickiest carpentry projects to build, even without adding any personality. I've probably built and hung four dozen gates for fences, sheds, coops, and pens — which has allowed me to discover and eliminate all the wrong ways to create gates.

The good news is that when you eliminate the wrong way, you have to replace it with a great way. I hope you'll allow me to share some great tips on building gates for your hen pen.

To stop predators, the gate should be covered with the same fencing as that used around the pen. Gates are also an opportunity to add some personality.

Building your gate on a flat surface like a driveway or patio makes it easy to keep the pieces properly aligned. Using screws instead of nails and putting diagonal bracing at the corners will keep the gate from sagging.

All Squared Away

We've all had the misfortune of walking through gates that sag: their bottom corners drag on the ground; their latches don't line up; they look sadly cockeyed as you approach them. What they probably all have in common is a lack of diagonal support. No doubt the gates were squared up when first built, meaning that all four corners made sharp 90-degree angles. But the combined effects of gravity, the gate's weight, and the occasional child swinging on them cause gates to sag, or rack, as carpenters say.

A rectangular gate without proper support will rack into a rhomboid shape. Like a rectangle, a rhomboid's opposite sides are the same length, but its corners are not squared up. It's a rectangle tilted to one side, like a speeding school bus in a comic strip.

One way to prevent racking is to incorporate a diagonal piece of wood that goes from one corner of the gate to the other, dividing the rectangle into triangles; another is to use triangular braces in the corners. In construction, triangles are more stable than rectangles.

In my gate built of cedar branches, each corner has a short wooden brace that's at a 45-degree angle. It forms a triangle at each corner to stiffen the gate and keep it square. If you go this route, be sure to predrill a hole at each end of the brace so that a screw slips through easily and then bites into the frame of the gate.

Using a short diagonal strut at each corner will hold the gate square; or you can use a single diagonal strut instead. If you use just one strut, it should start at the bottom corner near the hinge post side of the gate, meeting the latch post side at either the top corner or the halfway point.

Either way, the weight of the gate (and whichever child is swinging on it today) will be carried down through the diagonal strut and into the base of the hinge post, keeping the gate square. If you ran the diagonal from the top of the hinge side and down to the bottom of the latch side, it would provide zero resistance to racking.

Fasten gate pieces together with screws in predrilled holes.

Parts of Gates: A Primer

One post is the hinge post, and the other is the latch post. The horizontal parts of the gate are called stringers. The vertical parts of the gate are called stiles (or sometimes pickets), and some people call the diagonals struts.

More Gate Tips

- Another reason a gate might sag has nothing to do with the gate itself. If the gate*post* isn't stable, the gate's weight can pull it out of plumb. Then the latch won't line up or the gate may bind against the post. Make sure the gatepost is well supported by a properly mixed batch of concrete in a deep-enough hole (see chapter 3 for details).

- Don't build your gate until the posts are in place. Once they are, build the gate to allow a ¾" gap on all four sides to keep it from binding.

- If practical, orient your pen gate so that it opens out: it will be easier to slam it shut if the hens make a run for it. A gate that opens in will likely bind up on mulch in the pen.

Adding the Fine Touches

You can't install an Anti-Predator Apron in front of the gate the way you have along the pen fence (see page 59). Instead, you could just lay flat on the ground a section of fencing that extends about 16 inches in front of the gate to keep predators from digging under it. But a gate deserves a little extra effort to make it visually attractive.

To keep critters from digging under your gate, I recommend installing a Not-Welcome Mat made of brick, flagstone, or concrete pavers. Do this before building your gate, to ensure that your gate will clear the Mat by ¾". Choose several bricks or a piece of flagstone to reach from gatepost to gatepost and extend about 12 inches outside the gate.

Installing hinges on a gate is similar to installing them on a door. Attach the hinges to the gate or door first, not the post, and take your time – you'll want to do this only once (see page 98).

Your final step is to attach a store-bought or homemade latch at a comfortable height (see page 99). Then pour yourself a cold one. You deserve it. Aside from building stairs and roofs, you've just accomplished one of the trickiest tasks in carpentry.

HOW TO MAKE IT

NOT-WELCOME MAT

Predators can't dig through pavers of brick, concrete, or flagstone. These are perfect materials for a Not-Welcome Mat in front of and under a gate.

1. Dig a shallow trench from post to post and extending 12" outside the gate, just deep enough to accommodate the pavers resting on a 1" layer of grit. When installed, your pavers should protrude just above ground level so they won't be covered with silt after every rain.

2. If you don't have sandy soil, spread a 1" layer of grit (or a similar material called screenings) so that your paving will sit smooth and level. A rubber mat helps protect your knees.

3. Smooth out the grit/screenings and tamp it down with your feet. Start installing paving, and check to make sure the finished height is right. If not, add or subtract grit/screenings.

4. If some pavers are a little higher than others, use a dead-blow hammer to set them lower. This furniture-maker's tool can be found in any hardware store. It's heavier (and more effective) than a rubber mallet because it has a metal core, but its plastic covering protects the pavers from damage.

5. When you're satisfied with the placement of your paving, sweep grit/screenings into the gaps to keep them from wobbling. Sweep away excess grit/screenings for a finished look.

GATE HINGES AND HANDLE

1. Drill pilot holes where you have marked screw locations. Use a bit with a diameter that matches the diameter of the shaft of the screws. That way the threads will bite into the wood, but the shaft won't cause the wood to split.

2. With the hinges installed, prop the gate into the opening with a ¾" gap on all four sides. Set some scraps of wood that are ¾" thick (say, 1 × 4s) under the gate to hold it up. A helper can hold the gate upright while you work.

3. Have a helper hold the gate in place to make sure the hinges will meet the gatepost properly. Determine locations for the screws and mark them with a pencil or the point of a screw.

4. With the hinges in place over the gatepost, drill pilot holes as you did on the gate.

5. Drive the gatepost screws into place.

6. Attach a store-bought or homemade handle at a comfortable height.

7. Near the handle, carefully install a store-bought gate latch that keeps the gate closed but is easy to open.

MAINTENANCE TIP

After the screws are driven into the gatepost, enjoy watching your gate swing freely, but resist the temptation to ride it like a merry-go-round. That's the best way to make a gate sag, and a gate that sags even slightly won't close properly.

After the automatic pop door opens, Buttercup ventures out for breakfast.

Automatic Pop Door for the Flock

When I was creating our own Hentopia, one common feature of chicken coops promised to interfere with my goal of enjoying two-week vacations: the pop door. This is the small door that allows hens to go out in the morning and back in at dusk. The pop door should be closed after the hens go in at night to provide one more layer of security against predators like raccoons, possums, weasels, and coyotes. Then it needs to be opened again for the chickens in the morning so they can eat, drink, forage, socialize, and dust-bathe.

The simplest version I've seen is to put a couple of hinges on the top of a gangplank and a hook at the bottom. After the girls go in for the evening, just lift the gangplank to cover the door opening and secure it with the hook. Then at daybreak, or at least before breakfast, head out and drop the gangplank to open the pop door. It's a good design. But that's still a twice-a-day chore that precludes going on vacation. Or even sleeping in.

Chris and I are compatible in many ways, one of them being that we both firmly believe in "early to bed and late to rise." Neither of us is inclined to get up at daybreak to let the hens out of their coop.

We considered three ways to resolve this quandary.

- **Forget the pop door.** In the old days when the backyard sufficed as the chicken yard, pop doors were necessary to keep predators out of the coop. But for modern henkeepers, if you're confident that the fencing of the hen pen is secure against predators, then you could actually forgo an operating pop door. It is, after all, a redundant layer of security if the pen is well made. An opening big enough for the hens to come and go freely would only need to be about 8 inches wide by 12 inches tall.

- **Build a chicken-activated door.** Iva Biggin has come up with a seesaw-type automatic chicken door that looks doable to me, and another person, Andrew Wells in England, appears to have figured it out, too. The concept is that when all the chickens have climbed on the roost, their combined weight shifts the roosting bar so that cables and pulleys lower the pop door. When the birds hop off the roost in the morning, it shifts again and opens the pop door. To learn more, see the References and Resource section.

- **Buy an automatic coop door.** This is the route we chose. There are several designs on the market, all of which cost $200 or more. Chris said she was more than happy to pay if it meant we could wake up in a leisurely way while listening to the happy clucking of hens in their pen.

Decision Points on Automatic Doors

If you decide to buy an automatic pop door, it will come with decent installation instructions and videos. Instead of duplicating the manufacturers' efforts, let me share my thoughts on the various decision points when it comes to choosing automatic doors.

- **How to power the door?** If you already have electric power going to your coop, you're all set. If not, you could run an extension cord from your house or your garage to the coop, but you're taking a chance on an electrical short during rainy weather, so make sure the cord is plugged into a GFCI outlet. That way, if there's a short in the cord, the outlet will turn off rather than starting a fire in your house. An electrician can switch out an old outlet for a GFCI. Another

> **Chris and I both firmly believe in "early to bed and late to rise." Neither of us is inclined to get up at daybreak to let the hens out of their coop.**

option is to install a solar panel to run your automatic door. A solar panel is also immune to power outages. There are many options for small solar panels online and some automatic coop door vendors sell them as well.

- **Should you get a door that opens on hinges** or one that is raised up and down like a window shade? We didn't look at very many reviews of coop doors. "How different could they be?" we thought. We picked one that had a good installation video on the manufacturer's website, and it has worked fine for five years. But because it is the kind that's pulled up by a thin line on a spool, I anticipate that at some point the line will break and be time-consuming to repair. If I had to do it over again, I think I would go with a door that swings open on hinges. Why? Because a chain is only as strong as its weakest link, and a thin line might be that weak link.

- **Keep bedding contained.** Whatever type of door you choose, if bedding gets in the doorway, the door won't close all the way. A raccoon that gets through the fencing could grab the exposed edge of the door and open it. We positioned a foot-long length of 1 × 4 on the floor inside the doorway to serve as a shield to keep the bedding from blocking the opening.

- **A timer is a no-brainer. . .** All the automatic doors that I'm aware of depend on the common timer that turns living room lamps on and off while you're away. You set the *off* and *on* times according to the instructions, plug the cord into the timer, and plug the timer into the outlet. You're all set to go on a road trip.

MAINTENANCE TIP

Automatic doors rely on timers. Since they are working nonstop, they do wear out and will need to be replaced, perhaps every two years or so.

- **. . . Unless the power goes out.** Then the door will stay in whatever position it's in — open or closed. If the power goes out overnight, then the chickens will stay sequestered in the coop until you either prop the pop door open, prop open the keeper door, or are able to reset the timer when the power returns. This is the one downfall of the automatic doors — and a good selling point for solar panels. If this happens while you are out of town, it may be tricky to get a neighbor to reset the timer, as the process is a bit counterintuitive. For our out-of-town trips, we sometimes turn the door off and leave it open so that the chickens can come and go freely. Since we have a predator-proof pen, we don't worry about their safety.

- **Eliminate tasty-looking electrical wires.** After you install the door according to the provided instructions, there may be some visible wires that would be tempting for the chickens to peck at. That was the case with ours. To be on the safe side, I screwed a scrap piece of plywood over the wires in a way that wouldn't interfere with the operation of the door. Because I used screws, I could always remove the plywood if I had to replace or work on the automatic door. But I don't have to worry about the chickens electrocuting themselves if they mistake a wire for a worm.

A Keeper-Access Door for Yourself

I wanted a coop door that was big enough for me to lean inside for cleaning and maintenance on any part of the coop. If the automatic door fails, you can leave the keeper-access door open to allow the chickens to come and go.

Some chicken-keepers use padlocks on their coop doors because they are afraid raccoons will outsmart a latch. But I figure our sturdy pen will keep predators away from the coop. If a wily one did get in, the fact that the coop is set 18" above the ground means it can't reach the simple latches I use. To be safe, and to help keep the plywood door from warping, I used two latches on the keeper-access door.

Project Notes

The ¾" plywood for the doors may be labeled as $^{23}/_{32}$" plywood (since it's roughly ⅔ inch thick). Pressure-treated plywood is fine to use but not necessary. There are thinner (hence lighter and cheaper) sheets of plywood, but they are likely to warp, making the doors hard to close and leaving gaps for predators.

Some chicken-keepers use padlocks on their coop doors because they are afraid raccoons will outsmart a latch. But I figure our sturdy pen will keep predators away from the coop.

HENKEEPER'S DOOR

WHAT YOU NEED

- Tape measure and pencil
- Sheet of ¾" plywood
- Speed Square
- 2' level
- Jigsaw
- Clamps
- Sawhorses

- 2 hinges with screws
- 2 latches with screws

Optional, for circular window:

- 8" flower pot or similar circular object
- Drill
- Metal snips, or pliers with cutting blades
- 10"×10" hardware cloth or other metal mesh
- Eight ½" screws or fence staples to attach mesh

WHAT TO DO

1. Measure the width of the opening for the door.

Roughly 2 feet by 2 feet is enough room for an adult to reach into the coop. **Tip:** Make sure your opening is rectangular with 90 degree corners; if not, measure the top edge, bottom edge, and sides, so you can re-create that shape when cutting your plywood.

2. Measure the height of the opening for the door.

Leave about a ¼" gap on all four sides where the door meets the walls, so it doesn't bind up. In damp weather, wood absorbs moisture and swells up, so you want to allow for that.

3. Mark the cut lines on the plywood.

On ¾" plywood, measure and mark the dimensions of the door(s). Use the Speed Square and level to mark cut lines that are straight and perpendicular to the edge of the plywood.

4. Cut out the door.

Clamp the plywood to the sawhorses. Carefully follow the cut lines with the jigsaw.

5. Center the pot (or other circular object) on the door.

Use the tape measure to determine the center-line of the door and lightly mark it with pencil. Then use your judgment to determine how high or low you want the window to be by moving the pot up or down that centerline.

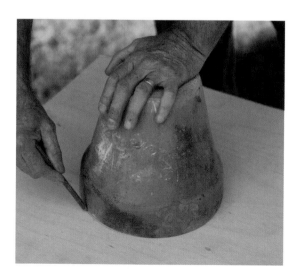

6. Draw the circle.

Holding the pot in place, trace its edge with the pencil. Remove the pot, stand up the door, and decide if you're happy with the window place-ment; if not, adjust it.

7. Make a hole for a place to start cutting.

Use a bit big enough to allow the jigsaw's blade to fit through easily. Cut a hole that follows the circular line.

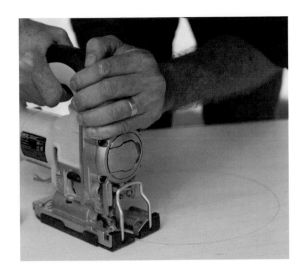

8. Carefully cut out the hole.

Insert the jigsaw blade into the hole. Make sure the blade isn't touching the wood until after you pull the trigger or the saw will kick back. Slowly follow the circular line. This can be tricky, so it's a good idea to practice a circular cut at least once on some scrap plywood. Best results will come from using two hands: one to guide the blade right down the pencil line, the other to gently swing the tail of the jigsaw out to keep the tool in alignment with the circle. Periodically blow away sawdust that may drift across the line, blocking your view.

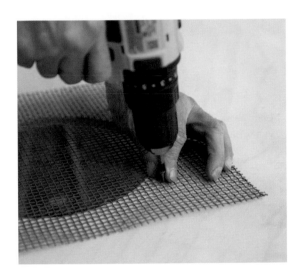

9. Install metal mesh or hardware cloth over window.

Use pliers or metal snips to cut mesh so that it will extend at least 2" beyond the window. Place mesh on the inside of the door, and use ½" screws or fence staples to attach it.

DOOR HINGES

Installing hinges is best done by two people: one to hold the door and the other to do everything else more easily.

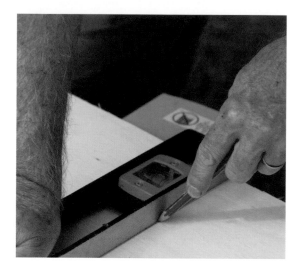

1. Mark positions for hinge screws on the door.

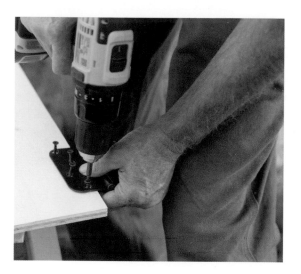

2. Attach the hinges to the door. Drill a hole using a bit that is the same diameter as the shaft of the supplied screws. That way, the threads will still bite into the wood, but the shaft won't split the wood.

3. Holding the door in place, mark the positions for the screws on the coop.

4. Screw the hinges into place.

HENKEEPER'S DOOR LATCHES

After the door is hung, install a pair of simple latches, about 3 inches from the top and bottom. If the eaves are wide enough to keep rain off, the walls and doors don't need painting. With a hinge and latch near each corner, keeper or cleanout doors are both secure and unlikely to warp.

WHAT YOU NEED

- Piece of 1 × 2 lumber
- 2 screws
- Drill with bits in various sizes
- Screwdriver

project

WHAT TO DO

Prepare the latch pieces.
1. Cut a pair of 3"-long pieces from the 1 × 2. Drill a hole through each piece about 1 inch from the end and just wider than the threads of the screws, so the latch will turn but not fall out.

Drill pilot holes in the wall.
2. Swap out the drill bit for one the same thickness as the shaft — but not the threads — of the screws. Drill two pilot holes near the edge of the wall that meets the door: one about 3 inches from the top, and one about 3 inches from the bottom.

Attach the latches to the wall.
3. Slip the screws through the latches and screw them into the pilot holes. Adjust the screws so that the latches hold the door snugly but can still be rotated out of the way.

Test run.
4. Turn the latches one way to hold the door shut. Turn them the other way to let the door swing open. Simple.

With doors the full width of the coop, there aren't any hard-to-clean corners. A rubber liner makes the floor easy to hose out, and extensions of the liner make it easy to sweep spent bedding into the compost bins.

Cleanout Doors

Cleanout doors are built just the same as the keeper-access door, with thick plywood and exterior-grade hinges. The differences are these:

- I recommend a pair of cleanout doors, big enough so that the entire width of the coop can be open for ease of cleaning out, without any hard-to-reach corners.

- I also recommend a pair of doors, so that each door can be aligned with a compost bin (more about this in chapter 6).

- Since a raccoon could stand on the bins, you want to use raccoon-proof latches, like hasps with carabiners.

  See pages 117–118 for more on cleanout doors.

6 No-Burnout Cleanout and Ventilation

We've talked about putting good things in the coop; now we must talk about getting not-so-good things out of the coop. This is where some people drop out of backyard chicken-keeping because conventional practices take too much time or are too messy.

But if you use the ideas described here, with no wheelbarrow to push, no trays to carry, no poop congealing on a wire floor, your coop will be clean, your hens happy, and you'll find plenty of time for enjoying cocktails and watching Hen-TV.

24/7

This may come as a surprise to you, but chickens are busy pooping 24/7. Yes, even in their sleep. Since they spend 8 out of 24 hours sleeping, about a third of their poop will show up in the coop, mostly right beneath their roosts.

That poop gives off moisture and ammonia. A modest amount of either is not a problem in the coop, especially if it is well ventilated; however, if either builds up, then your chickens may suffer respiratory illnesses that are hard to treat. For the health of your birds, you'll want to get the droppings out of there on a regular basis. But that "regular basis" doesn't have to be so often that it becomes a tiresome chore.

We rely on four components to manage the waste and the smell:

1. Proper bedding that absorbs moisture

2. Good ventilation that doesn't contribute to cold-weather drafts

3. An easily cleaned coop floor and wide cleanout doors

4. Accessible compost bins that make it easy to dump the poop out of the coop and eventually into the garden

For the health of your birds, you'll want to get the droppings out of there on a regular basis. But that "regular basis" doesn't have to be so often that it becomes a tiresome chore.

Proper Bedding

There are several choices for bedding, but they are not all created equal:

- **Shavings.** I recommend going with store-bought pine (not cedar) shavings, or dry sawdust from a sawmill or woodworker if you can get it. They both absorb moisture and are easy to sweep out when the time comes. Forgo cedar shavings — some poultry health people warn that the cedar oils are bad for both chicks and adults.

- **Sand.** Some people use sand, but since it's heavy and sticky when moist, it can be time-consuming to clean out of the coop. It's recommended to clean out sand every year, which sounds like hard work to me. And expensive over time. Others say it's dusty. Cheaper and lighter alternatives are available.

- **Pine needles or autumn leaves.** Free. Organic. Lightweight. Can't go wrong.

- **Straw.** Straw is fairly cheap and makes good bedding, but if mites get into your coop, they can hide in the hollow tubes of straw stems and become hard to get rid of.

Coffee chaff

- **Coffee chaff.** The brown, papery flakes that blow off green coffee beans when they are being roasted are called chaff. Coffee chaff from your local roaster makes a great bedding. The staff often bags it and gives it away to gardeners and chicken-keepers for free. The only downside is that coffee chaff will stick to the eggs in the nest box. Use pine shavings or some other material in the nest box if you don't want to lose time (and the shelf-stable nature of backyard eggs; see page 195) by washing chaff off your eggs.

Whatever bedding material you choose, a 4" layer should be enough to manage moisture in most coops.

So how often do you have to clean this stuff out? With seven hens, we scoop the poop from our coop maybe twice a year. Usually that's at the beginning of summer and the end of summer when we start to smell the ammonia. (Some people reduce the smell of ammonia by scattering charcoal in the coop, but I haven't tried that yet, so can't speak to it.) Every other year we completely clean out everything and add all-new bedding. A bale of pine shavings lasts a while that way. With more hens, you may need to clean things out more often.

Some folks fastidiously replace all the bedding frequently, but I think this is akin to dumping your cat's litter box rather than just scooping the poop. Even the most fastidious cat owners I know simply rake out the cat poop periodically and leave the litter in place until it can no longer absorb moisture. This is a good approach for chicken-coop bedding management, too. In any event, the ammonia smell and damp air will take a long time to build up if you have good ventilation.

How Does Bedding Work?

Moisture likes to travel from wherever-it-is to wherever-it-isn't. When the hens poop in the coop, the moisture in the poop migrates to the bedding. Then the moisture in the bedding migrates into the air flowing through the coop.

That's one reason why it's important to have good ventilation in the coop. You want that excess moisture to flow out, rather than build up and cause health problems for your birds. Also, flies can't successfully lay eggs in dry poop.

A Word about Deep-Layered Bedding

Some folks recommend something called deep-layered bedding. The idea is to complicate the construction of the coop by allowing a 12" layer of bedding on the floor. Then you throw some chicken feed in the bedding to encourage the birds to scratch around. This distributes the poop in the bedding and supposedly encourages the poop and bedding to compost. Color me dubious as to whether this is worth the additional effort and cost.

First, poop doesn't naturally compost inside the coop; rather, it dries out. To get poop to compost inside the coop, you'll have to add a measured amount of water. Too much, and the inside will get damp and moldy; not enough, and the poop won't compost. In fact, I don't recommend putting water in the coop.

Second, composting requires above-freezing temperatures. A compost bin benefits from the fact that the ground in winter is often warmer than the air. The compost bin will be much warmer than the bedding on the raised floor of a coop in the depths of winter: warm air rises and escapes a ventilated coop, and the cold air under a coop will extract any remaining warmth through the floor.

Using a deep litter method just allows you to put off cleaning the coop for a longer period by using a lot of pine shavings. Foot-deep bedding for a 32-square-foot coop (for 10 chickens) would require a small pickup-truck load of pine shavings.

My advice to all chicken-keepers is to keep the coop simple and the bedding about 4" deep, and to place the poop in a compost bin more than once a year.

My advice to all chicken-keepers is to keep the coop simple and the bedding about 4" deep.

Ventilation vs. Drafts

I'd say the most conflicted bit of advice in the poultry literature is the recommendation to provide ventilation but not drafts. Most sources don't explain the difference between the two things, much less how to have one without the other.

The difference hinges on the season. Good ventilation removes the ammonia smell and moist air from the coop year-round. In summer, drafts are a good thing: they remove hot air from the coop during the day and cool your hens' bodies at night. In winter, drafts of cold air will try to suck the heat right out of your birds: they all will need to eat more food just to stay warm, and older birds might get sick and die.

Although that sounds dire, remember that chickens have a thick layer of feathers, and at about 106°F (41°C), their normal body temperature is higher than ours. They have been surviving harsh winters in unheated coops in northern climates in Europe and Asia for a couple of thousand years. Your hens will likely survive winter even in a badly designed, drafty coop. But if you can sustain enough ventilation to move ammonia gases and excess moisture out in winter, without letting cold air blow over your chickens, then even your oldest and weakest chickens should survive the cold season.

For millennia, domesticated chickens around the world have survived harsh winter weather without needing a heat source in their coops. After all, they have down jackets.

Layer the Climate for Your Layers

Here's how it's done. Think of your coop as having three levels or zones from bottom to top:

1. The bottom level is about 15 to 18 inches high and has three components: the floor, the pop door, and the nest boxes. With the pop door closed at night, there should be no drafts here; install no windows or screens at this level.

2. The middle level is where the roosting bars are placed (see page 180). They can be from 15 to 24 inches above the floor. The roost level should have no drafts in winter. If you want to install operable windows for summer, this is a good height for them.

3. The third level is above the heads of the chickens. There should be plenty of ventilation here along the edges of the roof and in the gables (see chapter 10).

When you are constructing your coop, rather than covering the gables (the triangular spaces under both ends of the roof) with a solid siding like wood, I recommend just screening them with hardware cloth, regardless of climate. Predators will be locked out, and the constant cross-ventilation will draw out ammonia and moisture without creating cold drafts.

I also recommend filling the gaps between the roof rafters (where they rest on the walls) with chicken wire, which is more flexible than hardware cloth. This is best done before laying down the roof itself. Alternatively, use chicken wire to wrap the bottom ends of the rafters (where they protrude outside the walls). This can be done either before or after installing the roof. Cut the hardware cloth and chicken wire up to 1 inch wider than the openings they cover, and secure them with ¾" fence staples.

Our hens roost on small tree branches set about 18" above the floor. Chickens are not egalitarian, but with everyone on the same level there's less henpecking.

Cover any gaps between rafters, roof, and walls with small rolls of chicken wire. Hold the rolls in place with a couple of short screws. This allows hot, humid air to be vented out yet keeps critters from getting in.

To sum up: Ventilation at the bottom level? No. At the middle level? Maybe in summer, but not in winter. At the upper level? Yes, definitely, always.

Floors and Doors

Eventually you will have to clean the coop, so you'll want floors and doors that make the process easy. For the floor, the goal is to have a layer that won't let moisture soak into the wood. Wet wood can grow mold. You also want a surface that's smooth enough to sweep out easily and waterproof enough to allow you to hose down the floor on the rare occasion a total cleanout is warranted.

For the floor, the goal is to have a layer that won't let moisture soak into the wood.

Solution: A Waterproof Floor Liner

Some sources recommend using hardware cloth as a floor, based on the mistaken idea that all the poop will just magically fall through to the ground, where it will decompose. But chicken droppings will just stick to the hardware cloth and clog it up, creating a stinky mess. Any poop that does fall to the ground will not decompose, since it's too dry under the coop.

Other sources recommend installing a removable tray on the floor of the coop. That sounds like too much work to do and too many chances to decorate your clothes with last night's poop.

I prefer the relatively painless process of installing a waterproof liner on the coop floor. The best choice is a pond liner, which usually costs about $1 per square foot at a retail store. You may get one cheaper from a landscaper or nursery that installs ponds; a pond builder may even give you some scraps for cheap or for free. (You can also use plastic tarps or a heavy-duty vinyl shower curtain, but they may not last as long.) It doesn't matter what type of liner you use as long as it is waterproof.

To fasten the liner in place, spread a layer of exterior-grade adhesive on the coop floor and lay the liner on top. (Or cut the liner a bit oversize, bring the edges up the bottom of the walls, and staple it there.) Allow a "tongue" to extend out the cleanout door. When cleanout day arrives, the tongue can be positioned to bridge the gap between the doorway and the compost bins and function as a chute to channel the poop and spent bedding.

Positioning Doors for Easy Cleanout

When Chris and I need just to quickly scoop out some piles of poop and spent bedding, we can open one or both cleanout doors and position the pond-liner chute. With a plastic scoop, small shovel, or gloves (all of which we store on hooks inside the coop), we can gather the goods and drop them straight into the compost bins positioned next to the cleanout doors, and top up the coop with some fresh bedding.

When we need to do a more thorough expulsion of all the bedding, we open the keeper-access door as well as the two cleanout doors. From the keeper-access door, it's easy to use a push broom to expel everything out the cleanout doors, over the pond-liner chute, and into the compost bin. No cleaning hardware cloth, no lifting heavy trays, no wheelbarrow loading or unloading: the cleanout process is over before we know it.

No cleaning hardware cloth, no lifting heavy trays, no wheelbarrow loading or unloading: the cleanout process is over before we know it.

The matching coop cleanout doors are as wide as the wall. On the opposite wall, I have a keeper-access door (see chapter 5).

SPIFFY STEP-BY-STEP CLEANOUT

1. With a rake or push broom, push the old bedding out the doors into the compost bins.

2. The flaps of pond liner act as chutes to guide the spent bedding into the two compost bins.

3. With the compost bins closed, let the flaps hang down from the doorways so they're out of the way. Pour fresh bedding onto the liner.

4. Use a push broom or the flat side of a gravel rake to spread bedding about 4" thick. Coffee chaff, shown here, is free, organic, lightweight, and easy to spread quickly.

Compost Happens

Having chickens means having plenty of chicken manure for composting, a handy thing if you're a gardener as well as a chicken-keeper. You can forget, however, almost everything you've read or heard about composting. Much of the information out there is simply wrong or makes the process unnecessarily complicated. And almost none of the existing compost bins on the market are appropriate for backyard composting. If, like me, you want a compost system that (1) never needs turning, (2) never smells bad, (3) doesn't spawn fleets of flies, (4) doesn't attract vermin, and (5) delivers grade A compost for every spring and fall planting, then follow my Don'ts and Dos of Composting.

What if you don't need compost in your own yard? Let a neighbor harvest it in exchange for edibles and/or flowers or some other barter. Even if you can't make that kind of arrangement, composting will still reduce the volume of waste your coop produces as part of its natural process.

Save time and effort by putting a pair of compost bins next to the coop. Also store a large amount of "brown" material nearby and a small amount of garden soil that contains composting bacteria.

7 Building the Best Nest Box

Chickens will lay eggs in any number of places, convenient or inconvenient for you, safe or unsafe for the eggs. A well-built nest box will not only encourage them to lay those eggs in a secure spot but also make for easier collection on your part.

Nest Box Necessities

The best nest box meets three criteria: it is good-looking, easy to reach, and easy to operate. Ours became such a focal point of our yard that Chris (who in our household division of labor is the livestock manager) asked me (the facilities manager) to install a stepping-stone path leading up to it.

An exterior nest box with a hinged wall allows easy egg collection, even for children.

What We Wanted

As we prepared the nest box project, we put a lot of thought into the design so it would reflect all our concerns:

- **It had to be easy to access.** We didn't want to have to go through the pen, enter the coop, or get chicken poop on our shoes every time we harvested eggs.

- **It had to be the right height.** We wanted the neighborhood kids to feel they could help us by gathering eggs while we were away.

- **It had to be easy to clean** with a swipe or two of a whisk broom, which meant that it had to be arranged in a way that the hens couldn't poop all over it.

- **It had to be buildable from scrap pieces** to keep the costs down.

- **It had to be inviting** and cozy for the hens.

- **And it had to be cute enough** for us to look at every day.

What a Hen Wants

To answer our questions about the basics, we started with what a hen looks for in a nest box.

How many? A hen requires a nest box only long enough to plunk herself down and lay that day's egg. That means you don't need a separate box for every hen; in fact, you'll need only one box for every three to five hens.

Where to locate them? The hen wants someplace that's dry, dark, and out of sight of predators.

How big? She wants it big enough yet cozy. A one-foot cube open on one side works well. For bigger breeds, the box can be up to 14 inches on a side; for bantams, it can be as small as 8 inches. But many folks keep a variety of hens happy with all boxes 12 inches on all sides.

A nest box with a wall that opens as a hatch lets humans (especially kids!) meet chickens in a more controlled environment.

The exterior wall of the nest box is a cleanout hatch when it's hanging straight down. When resting on a swiveling support arm, it becomes a table for egg gathering.

Ins and Outs of an Exterior Nest Box

Many chicken-keepers mount nest boxes inside the coop, either set on the floor or attached to an inside wall. This is one valid option, with at least three downsides. The top of the nest box offers a surface on which chickens can roost and deposit poop all night (one more surface for you to clean). The coop must be big enough for you to enter (more time and money). Then when you do enter the coop to gather eggs, you get your shoes all poopy before you walk back to the house to cook an omelet (say no more).

Therefore, whoever came up with the idea of putting nest boxes on the *outside* of a chicken coop should get a medal. And a pension. And a monument on the National Mall. It's a great time-saver.

A nest box on an exterior wall of the coop — a wall that is also outside the pen — lets you pop out your back door to gather eggs in slippers, bare feet, or your best shoes without getting filthy. Chris and I both keep a pair of muck boots by the back door to wear on the rare occasions when we enter the pen to deal with the feeder or waterer, but we don't have to slip them on and off for our more frequent excursions to gather fresh eggs.

An exterior nest box also enabled us to build a smaller, less expensive chicken-scale coop. Think about the cumulative time and aggravation saved over years and years by not opening the coop door or pen gate ("Don't you girls sneak past me now!") each time you simply want to gather eggs. Finally, if the nest box is protruding from an exterior coop wall, the hens can't roost on it.

A nest box on an exterior wall of the coop lets you pop out your back door to gather eggs in slippers, bare feet, or your best shoes without getting filthy.

The Hatch

Before building our coop, Chris and I attended many coop tours and scoured many books and websites. When exterior nest boxes showed up, nearly all provided access through the roof of the box, which was hinged like a toolbox lid. Most of these henkeepers acknowledged that rain often leaked through the hinged edge of the roof lid where it met the wall of the coop.

One henkeeper, however, placed the hinges on the bottom edge of the outer wall of her nest box so that it opened like a breadbox. I still recall how excited I was when I saw this simple yet remarkable improvement. I've taken to calling that kind of hinged wall a hatch. (Appropriate for hens, no?)

This arrangement keeps the hatch secure against drafts and critters. When Chris and I want to collect eggs or clean the nest boxes, we have easy access and good visibility into the coop.

Closer view of the support arm, which swivels around a screw through its center.

SUPPORT SYSTEM

For the hatch to form a counterlike surface when it's open, you'll need a wooden arm that will swing out under it for support. I use scrap pieces of 2 × 2s, but any dimension will do.

Four Ways a Hatch Is Better than a Hinged Roof

The hatch has at least four great advantages over a hinged roof lid:

1. Rain doesn't leak in at the point where the nest-box roof meets the coop wall.

2. Kids and shorter henkeepers can see into the nest box to gather eggs, whereas they'd need a stepladder to reach down through a roof lid.

3. With a swiveling support arm that swings out from under the nest box floor (see facing page), this hatch also becomes a flat counter where you can set your egg carton while you gather eggs with both hands.

4. This arrangement also makes cleanup faster. With the hatch open and hanging down, use a whisk broom to sweep spent bedding straight out of the nest boxes and into a container, or onto the ground to decompose.

I used two arms on this nest box, in an overabundance of desire for symmetry. One support arm will do just fine.

With the support arm in place, the hatch can become a counter on which you can set your egg carton.

As a final touch, we dressed up the nest-box hatch with a drawer pull. It's merely ornamental, as it takes two hands to unlock the hasps and open the hatch, but it fits one of our design goals: it's cute.

The Right Height for You — and Your Helpers

We attached our nest box to the coop so that its bottom is hip-high for me and chest-high or head-high for neighborhood kids: in other words, the floors should be is 18" to 36" from the ground. At this height children can help sweep the nest boxes clean, gather eggs unassisted, or even meet a chicken face to face.

If you're at the open nest box long enough, one or more hens will come into the coop and then into the nest box to investigate the possibility of treats. This can be a good way to introduce a skittish child to chickens. An adult can easily close the hatch if the child gets scared — a much more manageable scenario than bringing a youngster into the chaos of a pen full of chickens.

Persuading the Hens to Accept It

Hens, even the most cooperative, may need a little encouragement to start laying in the nest box, no matter how well it is designed. Persuade them by parking something egglike in the box (a plastic Easter egg, a golf ball, or some such). This will tell the hens that some other, smarter hen thinks your nest box is a safe place for laying eggs, and they'll want to lay there, too. Chickens have a strong culture of follow-the-leader. Sometimes you have to be that leader.

Chickens have a strong culture of follow-the-leader. Sometimes you have to be that leader.

Fake eggs will help "prime the pump." To a chicken this looks like a good place to lay eggs.

Tools and Materials

Our nest box is built of plywood that is ¾ inch thick. You can use thicker wood, such as 2 × 4s, but I wouldn't go thinner than ¾ inch. You need that much thickness to minimize twisting as the wood dries, and to allow you to set a screw securely into the edge of each piece.

Make your cuts with a circular saw if you want to be fast, a table saw if you want to be accurate, a jigsaw if you want to be quiet, or a handsaw if you want to get strong.

FASTENERS

Screws will hold things together better than nails. Nails can pull out due to the parts they hold twisting, or from the shrinking and swelling of damp wood. But screws will hold their place because the threads are deeply embedded. And if you need to move the coop or want to enhance the nest box, screws will let you take it apart without butchering it.

You will use a pencil to mark where the screw will go through the first piece of wood, then predrill a hole that's a tad wider than the threads of the screw (see steps 3 to 5, page 135). The screw should slide easily through the first piece of wood and bite solidly into the second piece.

A jigsaw is a non-carpenter's best choice for a safe wood-cutting tool.

ROOFING

Since the nest box protrudes from the wall of the coop, it will need its own waterproof roof. I used a piece of shiny red scrap metal, but other roofing options will work too: asphalt shingles, cedar shingles, old license plates, flattened number 10 cans, a miniature "green roof," and so on. Think of the nest-box roof as small-scale but highly visible opportunity to dress up the coop and give it some charm and personality. If the roofing material on the main roof is attractive, there's no reason not to use the same thing on the nest-box roof.

A bright-colored roof draws attention to an exterior nest box.

HINGES

The hatch for our nest box has hinges at the bottom and latches on the sides. You could use gate hinges from the hardware store; choose ones that are made for outdoor use and won't get rusty. Make sure the screws aren't so long that they poke inside the box.

Country Hinges

I saved a little money by making three "country" hinges from brass screws and a scrap sheet of copper. (Other screws may cause copper to corrode.) With scrap sheet metal of any kind, predrill a hole in the metal that is wider than the screw's threads. Then mark and predrill a hole in the wood only as wide as the screw's shaft so that everything will be snug. These hinges don't move as smoothly as a gate hinge, but they are cheaper and work well enough.

It wouldn't work well for a door, but for a hatch that simply drops down, a strip of sheet metal will function like a hinge.

A hasp with a carabiner demands two steps to open: squeezing the carabiner and opening the hasp. That will keep raccoons from breaking into your nest box.

CRITTERPROOF LATCHES

When deciding on the latches for the external nest box on our coop, I tried to answer the henkeeper's eternal question: "Are these latches going to be safe enough?" That can also be translated as "What will my wife do to me if critters get her chickens?"

I checked the various chicken chat rooms online and found that a lot of people worried about raccoons getting into their coop. Some had lost so many hens that they were using combination locks to secure the doors. (I guess having that black mask doesn't qualify a raccoon as a safecracker.)

I think carabiners are tricky enough to keep raccoons out. I put a store-bought metal hasp and carabiner on each end of our nest box door to keep it secure and snug enough to minimize drafts.

Attaching Hasps

To attach the hasps, you may want a helper. One person holds the hatch in place, and the other puts the hasps in a convenient location. With a pencil, mark the location for the screws. Predrill the holes with a bit that is the same thickness as the shaft of the screw. That way the screw will slide smoothly through the holes in the hasp and the threads of the screw will bite soundly into the wood.

Swiveling safety hasps also require two steps to get them open, which should confound raccoons. And they take less time to open than a carabiner/hasp combination, so I may switch them out in the future.

A helper will be handy when you build this nest box.

HOW TO MAKE IT

THE BEST NEST BOX

WHAT YOU NEED

- 4'×4' sheet of ¾" plywood

- Tape measure

- Carpenter's square (I recommend a Speed Square)

- 2- or 4-foot-long level

- Marker

- Jigsaw

- Drill with assorted bits

- Screwdriver

- One box of 1⅝" exterior-grade screws

- One pair of 4" hinges

- Pencil

- One pair of 2½" latches

- 2"×2" scrap piece of wood about 10" long

- 2"-long screw to serve as support arm pivot

- Six 3" exterior-grade screws

- One 26"-long by 15"-wide piece of rolled asphalt roofing

- Utility knife

- One dozen galvanized roofing nails, ½" or ⅝" if you can get them, or ¾" if you can't

- Needle-nose pliers

WHAT TO DO

Prepare the plywood pieces.

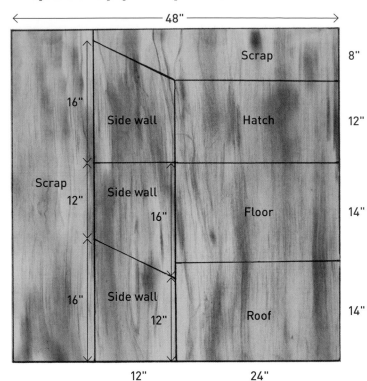

1. Copy this layout onto the plywood. Use a tape measure to measure dimensions. Use the Speed Square and the level to draw straight lines perpendicular to the edges of the plywood. Use the level to mark the diagonal lines of the walls.

Cut the pieces.

2. Make the long straight cuts or (as I recommend) have your local building store's trained staff do them. With a jigsaw, make the diagonal cuts at the top of each wall.

Assemble the floor, walls, and roof.

3. Temporarily stand the walls on the floor to mark their footprint. Predrill holes through the floor. Drill holes that are slightly larger than the diameter of the threads of the screws.

Note: The front edges of the walls meet the front edge of the floor. Double-check the drawing on page 137 before marking and drilling the roof and floor.

4. Turn the roof upside down and temporarily stand the walls on the roof, to mark the footprint. Predrill holes in the roof.

Note: The back edges of the walls meet the back edge of the roof.

5. Screw the roof and the floor onto the walls. You may be able to drive the screws without predrilling holes in the walls.

Attach the hinges.

6. Holding the hatch in place, position the two hinges about 2" or 3" in from the outer walls. Mark the holes with a pencil.

7. Predrill holes that match the shaft diameter of the screws (provided with the hinges) so the threads will bite into the floor and the hatch. Screw the hinges into place.

Attach the latches.

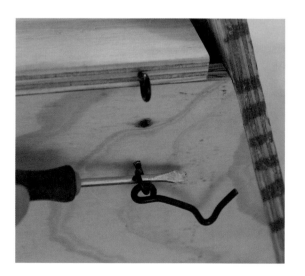

8. Position a latch (a hatch latch!) at each top corner of the hatch, and mark and predrill holes to match the diameter of the screw shafts.

9. Screw the hook and eye into place.

Attach the pivoting support arm.

10. At the center of the piece, drill a hole large enough to let your 2"-long pivot screw slip through.

11. Slide the screw through the support arm, and screw it up into the floor of the nest box, but not so tight as to keep the arm from rotating. When the arm is put away, it should be completely under the nest box.

Mount the nest box.

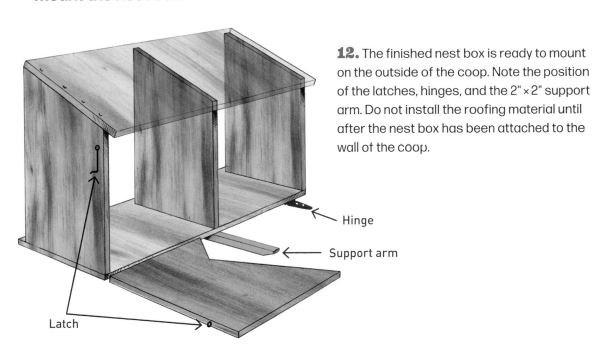

Hinge

Support arm

Latch

12. The finished nest box is ready to mount on the outside of the coop. Note the position of the latches, hinges, and the 2" × 2" support arm. Do not install the roofing material until after the nest box has been attached to the wall of the coop.

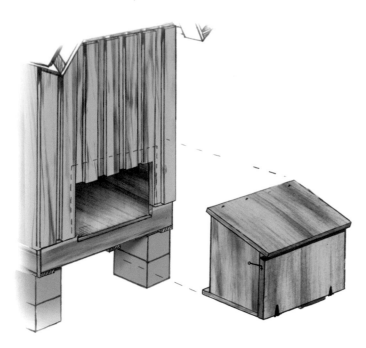

Attach the nest box to the coop.

13. Choose a spot for the nest box on an exterior coop wall, close to your back door for easy access.

Use a drill and jigsaw to cut out an opening that's 24" wide and approximately 12" high, that will put the nest box floor at approximately 18"–36" above ground.

Put the nest box against the opening. Slide the projecting flange of its floor in over the floor of the coop. Have a helper hold the box in place while you proceed through step 14.

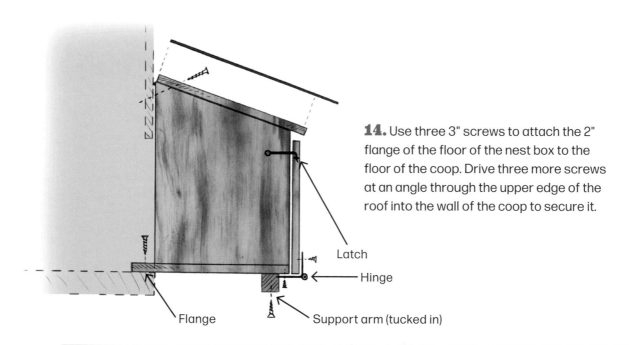

Latch

Hinge

Flange

Support arm (tucked in)

14. Use three 3" screws to attach the 2" flange of the floor of the nest box to the floor of the coop. Drive three more screws at an angle through the upper edge of the roof into the wall of the coop to secure it.

Making It Work

To open the hatch, just swing the arm out 90 degrees, pop off the latches and gently swing the hatch down to rest on the support arm.

Prepare roofing to fit on the plywood roof.

15. Mark a section of rolled roofing to 26" long and 15" wide. Using a sharp utility knife and a straight edge, cut out this section. To make the cutting easier and safer, put a scrap piece of lumber or plywood under the roofing.

Install asphalt roofing.

16. Position the section of roofing to allow a 1" overhang on the front and each side. The higher edge of the roofing should completely cover the higher edge of the nest box's plywood roof and meet the coop wall along its entire length.

17. With a clamp or helper holding the roofing in place, nail it down. You can hold each nail with needle-nose pliers to help get it started.

Award-Winning Chef Suvir Saran's Heritage Breed Chickens

Michelin-star chef Suvir Saran holds one of his Polish chickens in the door to his coop.

Suvir Saran grew up in New Delhi, which, he says, is "bigger than New York City in pop and hustle." As a child, he wanted to keep chickens, but his mother wouldn't let him. She said there was nowhere in their city home for them to free-range. Instead, he would pick up injured birds and nurse them back to health at home. Within a week, many of them would heal and be returned to the wild. He didn't know how to plaster their broken wings, but he used gauze and Scotch tape to put their limbs back in order. He fed them milk, grains, and packaged electrolytes. A chef taught him how to disinfect and bathe the birds in a solution of turmeric.

Journey to the Dream

Since coming to America in 1993, Suvir Saran has written three cookbooks and was the founding chef of the only Indian restaurant in the United States with a Michelin star, Devi in Manhattan. His lifelong desire to take care of chickens and other animals prepared him for taking care of friends, diners, and his American Masala Farm, near the Hudson River four hours north of New York City.

When he and his partner, Charlie Burd, bought the 70-acre farm in 2006, Suvir was finally able to realize his desire to keep chickens. He says that while most people keep chickens for practical purposes, he loves them for "idealistic reasons": their plumage, their cautious curiosity, the way the chicks follow the mother hen.

Heritage Breeds

Suvir and Charlie keep approximately 30 heritage breeds from the Livestock Conservancy watch list. The first ones were bought through McMurray's hatchery, but now Suvir buys them on EggBid.net, sort of an eBay for chickens. People bid against each other for rare breeds.

The farm's chicken population has ranged from 90 to 120 birds. Since these are heritage breeds, rather than hens bred for production, the flock may lay at most 40 eggs a day. "They are lazy birds, not the best producers," Suvir says. But he also claims they are "great performers and have great personalities," and help keep him sane and happy. He has named a few, such as a Crèvecoeur hen called Tina Turner and a Polish hen named Celine Dion.

If he had to choose a favorite breed, however, it might be the rare Penedesencas who, unlike most hens with white ear lobes, don't lay white eggs. When the pullets are young, their eggs may be nearly black. For Suvir they lay "the deepest dark chocolate eggs." A friend gave him one and after he saw the eggs he had to order more.

Belgian d'Uccles, also favorites, are affectionate and love to be with humans. "They could very well be housepets," he says.

Predator Patrol

Suvir hasn't had much trouble with diseases in his flock, but living in the country, predators are an ongoing threat: coyotes, foxes, fishers (similar to martens), bobcats, and minks. Once an apple tree fell and disabled the electric fence long enough for coyotes to get in and kill a baby goat. Suvir and Charlie repaired the fence but realized they had inadvertently trapped a couple of coyotes on the farm. They were both generally opposed to firearms, but Charlie, who grew up in West Virginia, procured a rifle and dispatched the coyotes with a few rounds.

They lost 60 birds to a mink one night when a caretaker left a tiny coop door open. After that they picked up some llamas and alpacas, which have been wonderful guard animals. They now have 25 of these fleece-bearing Andean animals.

The chicken eggs, along with the alpaca fleece, provide a small side business for Suvir and Charlie. Local restaurants pay $3.50 a dozen and the public pays $5 a dozen for the eggs. Given the cost of organic feed, Suvir says the income doesn't fully cover the time or effort.

Rooster Dramas

Suvir loves the plumage of roosters, but he hasn't been able to keep them for long. To minimize their pecking on hens, his roosters live outside the coop and eventually get taken by critters. The guinea hens managed to kill one rooster, so he keeps them separate now. One rooster was smart enough to cohabit with the alpacas and roost safely in their barn. Even he lasted only three years.

Suvir Saran's Cookbooks

In 1993 people started coming to Suvir's student apartment to enjoy his Indian home cooking. That led to catering birthday parties and even a dinner for Vice President Al Gore. He says, "One thing led to another and I gave up my student life as a designer, became a chef, cooking teacher, restaurant owner, and cookbook author." He met his partner, Charlie Burd, when Charlie attended one of Suvir's cooking classes.

His titles:

Indian Home Cooking: A Fresh Introduction to Indian Food by Suvir Saran and Stephanie Lyness

American Masala: 125 New Classics from My Home Kitchen by Suvir Saran and Raquel Pelzel

Masala Farm: Stories and Recipes from an Uncommon Life in the Country by Suvir Saran with Raquel Pelzel and Charlie Burd

Suvir Saran's Chosen Breeds

Ancona

Andalusian

Araucana

Belgian D'Uccle

Black Astralorp

Brown-Spotted Buttercup

Buckeye

Chantecler

Crèvecoeur

Cuckoo Marans

Delaware

Dominique

Dorking

Houdan

Jersey Giant

Lakenvelder

Minorca

New Hampshire Red

Orpington

Penedesenca

Phoenix

Polish

Rhode Island Red

Sebright

Silver-Spangled Hamburg

Wyandotte

Inside the Coop

When doors are closed the coop is quite secure and has some appealing features. For summer ventilation, the ceiling is even higher than that in the farmhouse, and the gables have wire covers for air flow. Windows also have wire covers for summer ventilation and can be closed like storm windows in long winters so hens stay warm from their body heat. There is a frostproof water spigot in the center of the coop so waterers can be easily refilled year-round. A couple of big skylights in the roof flood the coop with sunlight as a disinfectant and an encouragement to lay.

A people door opens into the center space of the coop, from which the owners can gather eggs from nesting boxes without disturbing the hens. One area is fenced as a nursery for newly hatched chicks. There, they are safe from bullying and Suvir and Charlie can handle and pet them, and, Suvir says, this "allows us to spoil them a bit."

When the chicks are nearly full size, they get access to the rest of the coop. In his cookbook *Masala Farm*, Suvir says, "We introduce them during the night as that is when the birds are at their most docile. You see, the minute the sun sets, chickens pretend that the world has ended — they just drop! We open the coop door, bring the younger birds in, and leave them on a perch. In the morning, they all wake up together and think they have lived together their whole lives. It's a beautiful thing."

The coop has three pop-up doors that let Suvir and Charlie control which fenced part of the farm the hens can access. Between growing seasons, the hens are allowed access to the vegetable garden to feast on its remains and till and fertilize the beds. During the growing season, the hens have access to the larger farm with its pasture, ponds, and 50 bushes producing blueberries, strawberries, raspberries, blackberries, gooseberries, and elderberries.

The hens aren't limited to layer feed and foraging. Because they host frequent holiday events and gatherings, Suvir says, "With a revolving door of visitors to the farm, there is never a dearth of kitchen scraps. They especially love citrus and chilies. Hot peppers are a favorite. In the winter, we make it a point to make extra soups and stews, vegetarian of course, that they get piping hot, and they savor them with glee."

Suvir would love to expand his flock to include a collection of bantam hens. For that he plans to build a mini-coop to keep them safe at night. He says, "The few bantams we have had, have been the most darling birds, and with the sweetest and most friendly demeanor. I am smitten by them."

Deviled Eggs with Cilantro, Chilies, and Spices

Suvir's spiced twist on classic deviled eggs. Even if the eggs are fresh, refrigerate them for seven days before boiling. Aging allows the inner membrane between the shell and the white to loosen, making the cooked eggs easier to peel. From *Masala Farm: Stories and Recipes from an Uncommon Life in the Country*

12	hard-boiled eggs
2	tablespoons peanut or canola oil
30	fresh or 45 frozen curry leaves, finely chopped
1	teaspoon brown mustard seeds
¾	teaspoon cumin seeds
1	small red onion, finely chopped
1½	teaspoons kosher salt
2	tablespoons of water, plus ¼ cup water
1	small jalapeño (remove seeds and ribs), finely chopped
1	tablespoon curry powder or garam masala
½	cup crème fraiche
¼	cup mayonnaise
¼	cup fresh lemon juice
½	cup finely chopped fresh cilantro

1. Halve the eggs; put the yolks in a bowl and the whites on a platter.

2. Heat the oil, curry leaves, mustard seeds, and cumin seeds in large frying pan over medium-high heat, stirring often, until the curry leaves are fragrant and the mustard seeds begin to pop, 1½ to 2 minutes.

3. Add the onion and salt, stirring often until the onion is browned on the edges.

4. Add the 2 tablespoons of water, and continue cooking until the onion is very dark brown.

5. Stir in the ¼ cup water, stirring and scraping any browned bits off the bottom of the pan.

6. Stir in the jalapeño and the curry powder (or garam masala) and cook until the mixture begins to stick to the bottom of the pan, about 30 seconds, stirring often.

7. Scrape the mixture into the bowl with the egg yolks.

8. Mash the yolks and spice mixture until very smooth.

9. Stir in the crème fraiche and mayonnaise completely.

10. Mix in the lemon juice and cilantro, and taste for seasoning.

11. Cover with plastic wrap and cool in the fridge for at least 1 hour.

12. Spoon a generous amount into each egg-white half, and chill or serve immediately.

PART III
BACKYARD BUILDING

There are few joys in life to rival that of building a sound structure that's attractive and functional. Here's your chance to experience that joy. You don't have to be a full-fledged carpenter to build a backyard Hentopia, and you don't need a big bank account, but you will need some knowledge. After 40 years of construction experience on two continents, I can simplify the process so that non-carpenters and the non-wealthy can create their own sound, attractive, and functional backyard coops.

8 Tips on Tools for Non-Carpenters

Let's not mince words. Power tools are potentially dangerous. Even hand tools like hammers can be scary for non-carpenters: you might flatten a thumb when you swing and miss the nail head. But it doesn't have to be that way. In this chapter I suggest an array of power tools and hand tools that pose fewer hazards, make less noise, and throw less sawdust than the carpenter's usual arsenal of claw hammers and circular saws.

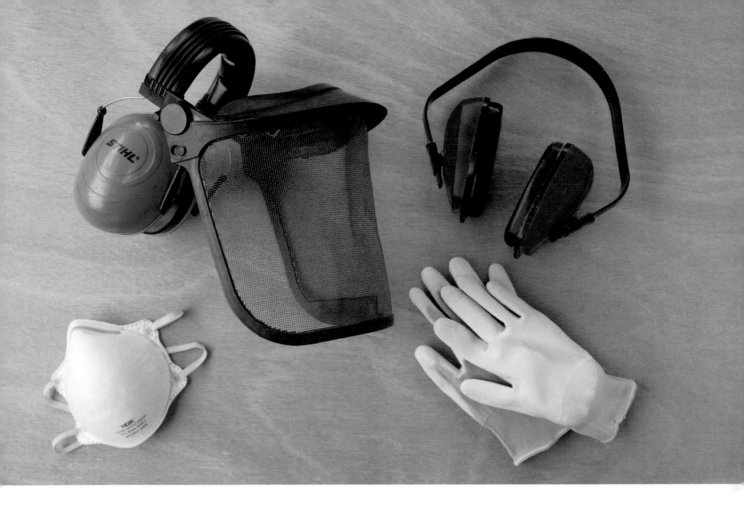

Safety Gear

Keeping chickens is fun, but it's not worth getting injured. Take precautions, stay focused, make safety the number one priority, and you will complete a great project without any regrets.

Even for professionals, any phase of construction is potentially dangerous. I remind all my helpers that working safely is the priority and getting the work done is secondary. Safety gear, a clean worksite, safe work habits, and a clear head are minimum requirements for all projects. Please don't try anything you read about in this book unless you're certain that it is safe for you. Remember your ABCs: Always Be Careful during your construction projects.

Gloves

Working with wood guarantees you'll get splinters. Working with metal roofing means you may get some cuts on your hands. Leather work gloves will protect you, but they are rather stiff and thick, which makes it hard to use your fingers for fine work like marking with a pencil or adjusting a tool. And they're hot in summer, when your body tries to

Better safe than sorry. Protect your hands, lungs, hearing, and face with proper gloves, face mask, ear muffs, and safety glasses or a face shield.

throw off excess heat from your feet, hands, and head. Shoes and gloves make it hard for you to stay cool. Thus I also recommend using a kind of glove common in gardening and in masonry work: cotton with a latex coating on the palm and fingers. They're thick enough to protect you, but thin enough for detail work.

Every few years I buy a bag of ten pairs of these gloves for $30 from my local hardware store or a cement products supply house. (You don't have to be a professional to shop at these places.) I stash pairs in each vehicle, tool drawer, or toolbox so they're always around when I need them. The coating of latex protects you from splinters and cuts, while the cotton breathes and lets some heat escape in summer.

Hearing and Face Protection

If you're working with power tools, even the quieter ones like jigsaws and drills, the noise can take a toll on your hearing. Operating a jackhammer for six months as a young man without hearing protection cost me a small degree of my hearing in my right ear. I won't let that happen again.

Don't skimp on hearing protection. Buy good-quality earmuff-style protectors that are rated to at least 20 decibels. Spending a little more also gets you a more comfortable fit. You're likely to wear them more often if they feel good. If you also want something to protect your face from flying sawdust and debris, look for a chainsaw dealer and buy their hearing protectors combined with a screened face shield. The screen doesn't fog up like a plastic one does, and it allows airflow to keep your face cool. You may have to special-order it to get the one without the attached hard hat.

If you don't have the screen face shield and if you don't already wear glasses, wear safety glasses to protect your eyes from flying debris. Choose a pair with scratch-resistant polycarbonate lenses.

Don't skimp on hearing protection. Buy good-quality earmuff-style protectors that are rated to at least 20 decibels.

See Saw, Make Cuts

A circular saw can be a scary tool to operate. I don't recommend using one without getting close guidance from an experienced carpenter or woodworker. Many, if not most, of the cuts for building chicken habitat can be made with a much quieter, less dusty, and easier-to-handle power tool called a jigsaw. (I still recommend getting some close guidance if this tool is new to you.)

Most professionals use a jigsaw and its narrow blade only for cutting a curved line in a piece of wood. Despite being slower than a circular

saw, a jigsaw can also be used for straight cuts on the 2×2s, 2×4s, and pieces of plywood that compose most of the components for a chicken habitat. It can even cut a 4×4 for a post if you buy cutting blades that are longer than those that normally come with the tool.

To use a jigsaw, simply make a pencil mark where you want to cut the work piece. Clamp the work piece to a couple of sawhorses so that the cut portion will simply fall off. With both hands on the tool, rest the front edge of the flat shoe on the edge of the wood. *Don't let the blade touch the wood until after you've started the tool or it will kick back.* Guide the tool with both hands as you make the cut.

Once you've mastered making straight cuts with your jigsaw, you may want to use it to make curving cuts for things like circular windows in plywood walls. Even for professionals it's tricky to get a jigsaw to perfectly follow a curving line, so you may want to smooth out the curve with a power sander. Or you may just remind yourself that you're building chicken coops and nest boxes, not heirloom furniture. I figure that if it looks good at arm's length or better, it is good.

The narrow blade of a jigsaw not only allows the user to make curving cuts but also throws off less sawdust than a circular saw and is less hazardous.

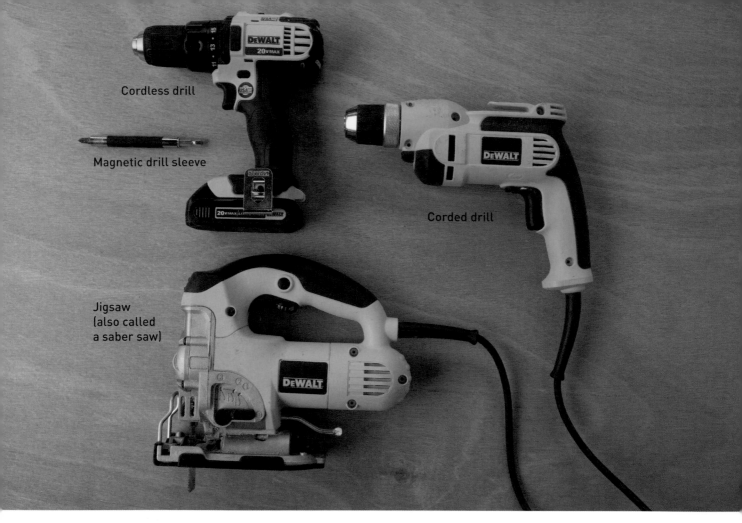

Cordless drill

Magnetic drill sleeve

Corded drill

Jigsaw
(also called
a saber saw)

Power Drills

Screws, when installed correctly, hold much better than nails. And driving them home with a drill and an appropriate bit is a lot easier and safer for a novice than pounding with a hammer and risking your thumb. Using a power drill and screws has a third advantage: if you want to take your structure apart or make changes, you can easily back the screws out and start over.

Predrilling

The best practice if you're installing a hinge, latch, or handle is to predrill. Find a drill bit with the same diameter as the shaft of the screw, but not as wide as the threads, and drill the hole. Then use a screwdriver or, better yet, a power drill with a screwdriver bit to quickly drive the screw into place without bogging down or splitting the wood. The threads of the screw will bite snugly into the wood, holding the fitting in place for a very long time.

When driving a screw through one piece of wood and into a second piece, as you might when attaching a floorboard to a floor joist, I'd recommend a slightly different predrilling practice. Find a drill bit with a diameter slightly greater than the diameter of the threads of the screw but not as wide as the screw's head. Predrill a hole in the first piece of wood. The screw will fit easily through this first piece by hand. Then it's ready to be driven into the second piece of wood. With the screwhead snug against the first piece and the threads biting into the second piece, you'll have a very tight connection.

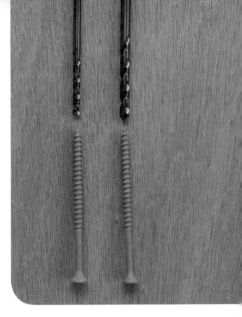

The drill bit at top left is the same diameter as the shaft of the screw at bottom left, so the threads will bite into the wood. The drill bit at top right is wider than the threads of the screw at bottom right, so the screw will pass through the first piece of wood.

Magnetic Drill Sleeves

I can't recommend magnetic drill sleeves enough. They only cost about $5, but they save a lot of hassle. They fit into any drill. You can slip a wide range of bits into the end. They're magnetic, so you're less likely to drop your screw. And the sleeve slides over the screw, which helps hold it in place until the threads bite the wood.

The sleeve also makes it easier to keep the bit and screw in alignment. If you've been stripping bits or screwheads, it isn't the fault of the tool. It's operator error. If the bit and the screw aren't in alignment, you will hear that horrible barking sound of the screwhead being stripped to the point where the bit won't drive it anymore. Stop as soon as you hear that sound, then take your time and get the bit aligned with the screw. The sleeve helps with that, but it does need a little guidance from you.

With the screwhead snug against the first piece and the threads biting into the second piece, you'll have a very tight connection.

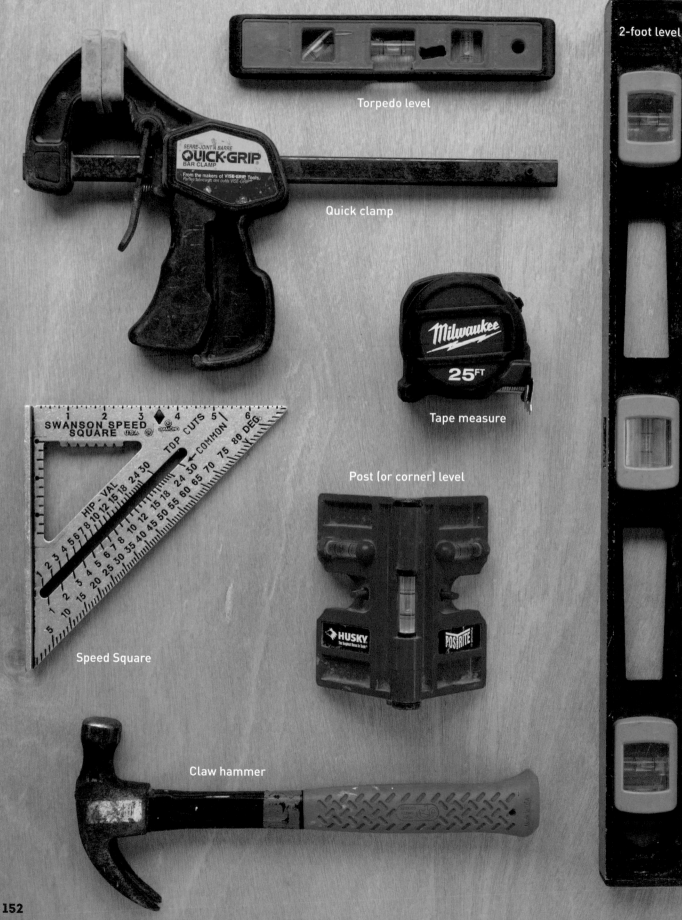

Torpedo level

2-foot level

Quick clamp

QUICK-GRIP
SERRE-JOINT · BARRE
BAR CLAMP
From the makers of VISE-GRIP Tools.
Par les fabricants des outils VISE-GRIP

Milwaukee
25FT

Tape measure

SWANSON SPEED SQUARE USA

Post (or corner) level

HUSKY
The Toughest Name in Tools

POSTRITE

Speed Square

Claw hammer

Another Option

Since most of the wood you'll be using for your coop and pen will be relatively soft, like pine and cedar (as opposed to much harder woods like oak and maple), you may be able to screw two pieces of wood together without predrilling at all, if your hands are strong enough. You and a drill may be able to drive a 2" screw straight through a sheet of ¾" plywood and into some framing, or a 3" screw through one pallet and into another. Pine and cedar are not likely to split unless your screw is too close to the edge. Give it a try on a couple of sample pieces of wood and see how it goes.

Hand Tools

Power tools get all the attention, but it's the smaller hand tools that can make a big difference both in how well the project goes and in how much job satisfaction you find. A backyard henkeeper will want to have certain hand tools at the ready for repairs, remodeling, or new construction.

Pliers

Pliers are handy for holding objects instead of holding them with your hand, such as short nails when you don't want to smash a thumb. Many of them also have a blade near the hinge for cutting thin wires, as on a fence, and they have grippy teeth that can help you turn a bolt or loosen a garden hose connection.

A backyard henkeeper will want to have certain hand tools at the ready.

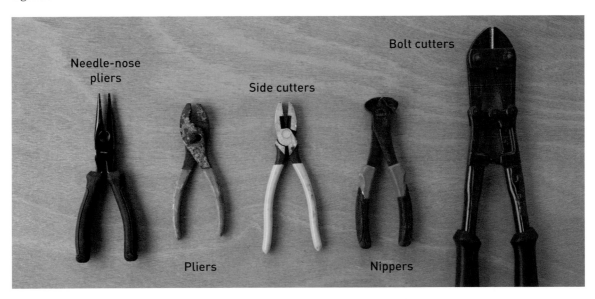

Needle-nose pliers

Side cutters

Bolt cutters

Pliers

Nippers

Speed Square

The Speed Square determines if a corner is square and can also be used to mark cut lines. It is one of my go-to tools. First off, it's indestructible, with no moving parts to fail. Second, it's the right size to fit snugly in my back pocket.

It's a flat, triangular piece of metal, and each of its three sides has a purpose. One side has marks for measuring or marking lengths. Another side has marks for reading angles. The third side has a lip so you can slap the Speed Square on a plank and mark a perfect 90-degree line for cutting a board (or check a new board to make sure the ends are square). You can also flip the lip and use it to draw a perfect 45-degree angle on a plank. Since a Speed Square is a right triangle (one corner is a 90-degree angle), you can use it to make sure corners of a raised bed or other structure are square, or that the blade on your circular saw is square with the shoe.

Why is it called a Speed Square (and why is it capitalized)? Carpenter Albert Swanson wanted a tool for laying out roof rafters that would be faster than a combination square or a roofing square. He invented this little gem in 1925, trademarked the name, and founded a company to produce it. And now, like Kleenex and Dumpster, the name has become so common that people often don't realize it's a trademark.

Claw Hammer

Look at the head of a carpenter's claw hammer, and you'll see it's actually two tools in one: the hammer end lets you drive a nail home, and the claw end lets you pull out all the nails you've bent. But sometimes a nail is so long or so tightly held that the claw only pulls the nail out partway. A good carpenter's trick is to put a piece of scrap wood under the claw to amplify your leverage.

Three-Way Plug

Most likely, you'll be using more than one power tool on a carpentry project — perhaps a jigsaw to cut planks or plywood, and a drill to fasten them together. If you're in the South in summer, you may also be running a big fan to stay cool. I know I do.

Switching plugs back and forth slows things down, and using more than one cord is an unnecessary expense. Invest instead in a three-way plug. Some carpenters use power strips, which allow for half a dozen tools, but office equipment on a job site just looks wrong to me. If you need more than three tools (a reciprocating saw, perhaps), put another three-way on the first one, allowing connections for five tools at a time.

I keep all my cords and three-ways together by storing them in a repurposed 15-gallon nursery pot with handles. You can probably get one for free from a local landscaper or garden center (see page 22).

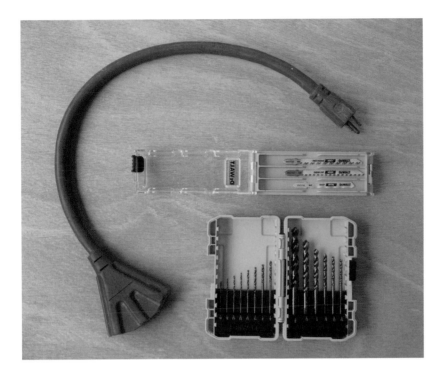

A three-way plug lets one cord power up to three tools. Jigsaw blades and drill bits are best stored in small cases so they don't go on walkabout.

Tape Measure

Almost all tape measures are disappointing in one way or another, which is strange given how necessary they are in almost any home, garden, or farm project. For a tool that should be reliable and handy, they are mostly a handful of small problems. Either the hook doesn't stay hooked, or the tape is too flimsy to extend for much distance, or the spring doesn't retract the tape without plenty of assistance, or it retracts so fast it hits your finger. Ouch. Then the tape or the spring breaks anyway, so you have to throw it away and buy another one. You spend even more money, hoping for a marginally better tape measure. But still they disappoint.

Finally, a company has invested in developing a tape measure that works really well and should last for a long time. Milwaukee Tools has a red and black tape measure that solves all those problems and costs only a little more than inferior tapes. It has a 360-degree hook so that you can pull a measurement off the most obtuse angle. The tape is stiff enough to extend 9 feet, unsupported. It extends and retracts easily, and the belt clip is designed so that it doesn't tear up the fabric of your pants pocket if that's where you carry it, as I do. It even has a stop to protect your finger when the tape retracts. It feels sturdy enough that you may be able to pass it down to the next generation.

Milwaukee offers many options: I recommend the nonmagnetic 16-foot tape for non-professional use.

Level

Also called plumb sticks by carpenters, levels are lightweight bars, usually either two or four feet long, embedded with a spirit level, a glass tube filled with a liquid that has a bubble in it. When the bubble lines up nicely between the two lines on the tube, that defines "level to the horizon." A level can also tell you if a post is plumb, another word for straight up and down. The bubble in the tube of liquid should appear between the two lines for something to be level or plumb.

You can also get by with what's called a "torpedo level," named that because it's pointed on each end. This small item, usually about 8" long, will fit easily into a pouch, sleeve on your tool belt, or back pocket.

Levels also help you mark straight lines.

Quick Clamp

Some projects absolutely require a helper, but in many cases a clamp can replace a human assistant. One of its best uses is to hold a work piece secure on your sawhorse while you cut it. Granted, it won't be as chatty as a helper. That may be good or bad.

Conventional clamps, however — called bar clamps and C-clamps — take two hands to attach or release. Quick clamps are much better. They can be installed and removed with one hand. Hold the pistol-grip handle and squeeze the trigger to tighten it up. When you're done, grab the handle and squeeze the release lever to loosen it. When you're working alone, a quick clamp can save lots of time and fiddling around.

Quick clamps can be operated with one hand.

Keeping Tools Handy

Think about your kitchen tools for a minute. Maybe, like Chris and me, you have spatulas, tongs, and wooden spoons in a container near the stove; a rack of knives near the cutting board; and sponges and detergents arrayed by the sink. Would it make any sense to put all those tools somewhere deep in the pantry? Of course not. Why, then, are your most commonly used chicken-keeping tools buried somewhere in the garage?

We have three tools that see a lot of use in the coop and pen, so we've given them homes that keep them handy:

Broom. When a hen mistakes her nest box for a toilet stall, it needs sweeping out. We took the rattiest whisk broom we had in the house and dedicated it to that task. I found an old teacup hook in the utility drawer of our kitchen and screwed it into a cedar post within arm's reach of the nest box. Since it's under the eaves of the coop's roof, the broom rarely gets wet, even if it hasn't grown any better-looking.

Chopping knife. Sometimes we feed kitchen scraps to the hens. As I toss the scraps into the pen I realize the birds will have an easier time with that cucumber or melon if I chop it in half. I keep a yard-sale chef's knife stuck blade-down into the soil just inside the pen gate. Yes, it's ratty, too, and rusty. But it's handy and it cuts quickly. In spring, the knife is redeployed to the asparagus bed. There it stands at attention with the blade tucked into the soil. That way I can cut fresh spears anytime I'm in the garden, without making a trip to the garage or kitchen first.

Rake. Every blue moon, the thick layer of wood chip mulch in the pen piles up in drifts that interfere with access to the feeder and waterer. I normally notice this only when I'm topping up one of them, so I would hate to have to turn around and go to the garage to scare up a gravel rake. I have taken our rattiest gravel rake and drilled a hole near the top end of the wooden handle. The hole is big enough to fit over a nail that I drove into a corner post of the coop, where the rake is handy for mulch smoothing and stays dry under the roof eave.

9 High-Quality, Low-Cost Materials

Martha Stewart I'm not. You don't have to be either. You don't have to spend a lot of money and time creating a quality coop and hen habitat. In this chapter I'll share suggestions on inexpensive materials and where to find them.

Coffee chaff is not only excellent chicken bedding but is also great mulch in vegetable beds. The burlap sacks that the coffee beans come in can be used as free mulch in the pathways.

Materials for the Girls

I've mentioned some of these earlier, but here's a chance to go into more detail.

Coffee Chaff

A large garbage bag full of lightweight coffee chaff (see page 113) will cover the floor of a 5' × 5' coop about 5 inches deep. Since the chaff has been exposed to high heat, it won't introduce any pests or diseases. Coffee chaff can also be used as bedding in worm bins (see page 32) and as the brown portion in compost bins (see page 119). It is also great as mulch in vegetable beds. Most large or midsize cities (and even some towns) have their own coffee roasters now, and these businesses generate lots of coffee chaff that they bag up and need to get rid of. Sometimes they pay a large-scale compost operation to take it off their hands, but often they pay to dump it in a landfill. It's very likely that your local roaster will happily let chicken-keepers and gardeners pick it up for free.

Wood Chips

As noted in chapter 4, wood chips make excellent mulch in the fenced area for chickens. Whenever arborists or landscape companies cut down trees, they often run the branches through a wood chipper that shoots the chips into a dump truck. This makes the branches more compact and easier to dispose of. They would like to avoid paying a fee to dump the wood chips at a landfill, so they are more than happy to dump the wood chips for free at your home. Do a web search of arborists in your area or speak to a crew when you see them working. Put a tarp down before they dump, though; it will make cleanup go a lot faster.

Welded Wire Fencing

You'll need rolls of welded wire fencing (see page 55) to build the pen and the gate, and to cover the vents in your coop to keep predators out. I found free vintage-style fencing to use in our pen at the site of a house remodeling. The workers had rolled up the old fencing and put it by the curb in preparation to take it to the dump. I saw it in the nick of time and asked if I could have it. The workers laughed as if I were some kind of fool for wanting their "trash." But visitors to our Hentopia always enviously comment on the unusual pattern.

You can, of course, buy fence wire from your local hardware store or big-box home improvement store, but you can save a lot of money by buying leftover rolls of wire at your local metal recycling facility. Do a Web search to find these places. Then call to determine whether they set aside valuable things for sale to the public. Many of them do, as they can get a better price that way (yet you'll still save half or more over retail). Be aware that it's a catch-as-catch-can sort of thing. Swing by or call periodically to see if they have any fence wire on hand, as it may sell quickly. This may also be a good place to get usable metal roofing, hinges, tchotchkes, and other metal parts for your coop and your hen habitat.

Pallets

Pallets are lightweight wooden platforms on which many products and appliances are stacked, strapped, and delivered all across the country. They're shaped so that a forklift can pick them up and move them. The majority get tossed in a landfill after delivery, but they are still useful.

The most common pallets are made of pine and are 44 inches long, 40 inches wide, and about 5 inches high. They can be used whole to build the frame for a coop's floor, walls, and roof. And you can break them down (see page 162) to use the individual planks for building nest boxes, doors, siding, and other parts of the coop.

Some vintage fencing has a V-shaped pattern rather than modern rectangles, squares, and hexagons.

Taking Apart a Pallet

The common way to take apart a pallet requires a lot of elbow grease: pull the planks apart with a pry bar, and then back each and every nail out with a hammer. Depending on your strength, it might take a good hour for each pallet. Dismembering three pallets for the siding on your pallet coop could take half a day with a hammer and crowbar.

With a metal-cutting blade and a reciprocating saw (Milwaukee Tools makes one called a Sawzall), a handy person can dismember a pallet in 10 to 15 minutes and have enough free lumber for the siding on a pallet coop in less than an hour.

The metal-cutting blade, which has tinier teeth than a woodcutting blade, can slip between the planks and cut the nails right through. You'll end up with three 44"-long 2×4's (good for roosting bars) and about a dozen 40"-long 1 × 4s and/or 1 × 6s for siding.

Here are a few handling tips for using power tools to break down the pallets:

- Wear earplugs or earmuffs to protect your hearing.

- Wear cotton or leather gloves to keep from getting splinters or snagging on a cut nail.

- To keep the saw from rattling the teeth out of your head, be sure to push the foot of the tool — the flat collar around the blade — against the wood to absorb the vibrations.

- To cut through the nails faster, gently rock the saw up and down while keeping its foot snug against the wood.

- Sometimes the blade will cut into the wood a little, but that's not a problem.

- Before each cut, check the teeth of the blade. If they're worn down, switch out the blade for a new one. A dozen pallets may use up two or three blades.

When finished, stack the lumber where it won't get wet. Put bricks under the stack to keep it off the ground and a tarp over the lumber to keep rain off until it's time to build the coop.

Not every business that uses pallets will sell or give them to the public, but many do, and often they're only a buck apiece. Companies that might give away pallets for free include garden centers, feed and seed stores, heating and air conditioning businesses, appliance stores, stone yards, grocery stores, big-box stores, hardware stores, and many kinds of industrial facilities. Drive through commercial and industrial parts of your town, and you will see stacks of pallets waiting to be disposed of.

A minority of pallets have been treated with insecticides such as methyl bromide. In most cases these pallets have been shipped from overseas and the chemicals are applied to keep insects from spreading to another country. Such a pallet will have the letters "MB" stamped on it. I would not recommend using such a pallet.

If a pallet has no such stamp on it, it is probably produced in the US for one-time use and not treated at all; thus, it is safe to use.

Other pallets may have the letters "HT" stamped on them, which stands for "heat treated," another, safer way to stop insects; they are also safe.

A metal-cutting blade on a reciprocating saw will cut all the nails on a pallet in less than 15 minutes.

Pressure-Treated Lumber

Although, as noted in chapter 3, no pressure-treated wood has been made with toxic arsenic and chromium since 2003, I still see recent books and websites by gardening and chicken-keeping experts who aren't aware of this. They warn people not to use pressure-treated lumber because of fears about toxins.

Pressure-treated lumber is now saturated with two safer products: a liquid solution of copper, the same material that the modern water pipes for your house are made from; and quaternary ammonium, the chemical that is in Formula 409 spray cleaner, which is made to be safely used on kitchen countertops around food and kids. Both treatments keep fungus, termites, and other soil-borne critters from decomposing any wood that touches the ground.

For your coop, you only need pressure-treated wood for any parts that are touching the soil. In most cases, that means the posts that hold the coop up off the ground or the posts for the pen. The roof will keep the rest of the coop dry enough so that it won't rot, just as with your house.

The parts of the pen that aren't touching the soil — such as gates and stringers that hold the fencing over the top of the pen — will last at least 10 years even if they aren't pressure-treated. But if you don't want to be replacing those rain-exposed parts every decade or so, then do use pressure-treated wood. It will cost more than regular lumber, but you probably won't have to replace it in your lifetime.

A 4" × 4" × 8' post of pressure-treated pine. The label says "Approved Use – Ground Contact," which means that it will resist rotting for about 40 years.

Rot-Resistant Posts

Some people may still want to forgo pressure-treated wood, and that's okay. There are some naturally rot-resistant species of wood: cedar, juniper, cypress, redwood, and black locust. Depending on soil types, rainfall, and quality of wood, they may last 10 to 20 years. Again, many experts recommend these woods without knowing that all parts of the tree are not rot-resistant. Under the bark of any tree you'll find several inches of light-colored material, called sapwood. This part of the tree is not rot-resistant at all.

At the center of the tree you'll find several inches of dark-colored material, called heartwood. That's the part of these species that resists termites and fungus. If you're buying posts, look at each end to make sure that all or most of the wood is the dark-colored heartwood. Same for planks: make sure most, if not all, of the plank is heartwood. Otherwise, the sapwood will rot as fast as the cheapest wood, but you'll have paid a premium price.

Stains versus Paints

We have not painted our coop. Chris likes to see the wood grain on the plywood exterior. With wide-enough eaves (ours are about 8 to 12 inches wide), you'll rarely have rain striking the walls or doors. We don't have to worry about the wood rotting or warping.

The ultraviolet (UV) radiation in sunshine *can* break down the fibers of wood. If you see a pressure-treated deck that looks shabby, it's not because of rot. It's because the UV radiation has damaged a lot of the wood fibers. Fortunately, our coop is on the north side of the garage, so it isn't exposed to much direct sunlight.

All that is to let you know that covering the wooden walls and doors of your coop with a paint or stain may not be really necessary. If you do want to color it (or protect the wood on a sunny site from UV radiation damage), then I recommend going with stain rather than paint. Both products function as what's called a "sacrifice layer" that the UV breaks down rather than the wood. As they break down, the paint will peel and look raggedy; the stain will only fade.

The other thing to know is that the only reliable way to differentiate the good, the bad, and the ugly among paints or stains is to check the price. The more expensive it is, the more pigment it has and the better coverage (and longer life) you'll get. So, if you don't want your time to be wasted, buy as expensive a stain as you can afford.

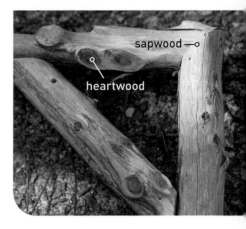

This gate is made of eastern red cedar. The dark wood visible in the top part is the rot-resistant heartwood, while the light-colored outer layer is sapwood, which can rot quickly when exposed to damp soil.

Places to Get Free or Low-Cost Supplies

We live in a rich country and lots of people dispose of things that are still useful. Fortunately, there are many outlets for items that you can repurpose.

Architectural Salvage Stores

Architectural salvage stores are retail establishments that scour construction sites and remodeling jobs to scoop up reusable materials like bricks, lumber, doors, fencing, and windows. Many contractors also bring materials to these sites to avoid paying landfill fees. You can find very pretty supplies here at a very good price compared to buying new. Plus, anything you buy will probably come with a story and some vintage cachet.

Lumberyards

If you want naturally rot-resistant lumber, search for a local lumberyard. These aren't as common as they used to be, so you may have to drive some distance to get to one. To find a lumberyard, ask around at a hardware store or a locally owned building supply store, or contact local woodworkers. Then call the yard first to see if they have the posts, planks, or species that you're looking for. Sometimes they can cut posts or planks to the dimensions you need. Not all of them deliver, so you may need to drive your pickup truck (or borrow one) to pick up your order.

I bought the cedar posts for our coop and pen from a lumberyard. They also had a big pile of cedar stumps at a very good price. I didn't know what I was going to do with the stumps, but I had to buy them, bring them home, and pile them behind the garage. Chris immediately saw their best use: "Stack them upside down like a totem pole," she said. I think I'll get to that this year. Or next.

Metal Scrapyards

Not all scrapyards allow sales to the general public, so call around first. If you get turned down over the phone, it's still worth a face-to-face visit. With a friendly chat and a mention of your chicken coop dreams, the owner may relent due to your charm and your amusing project. Laws shouldn't be broken, but some rules *can* be bent.

At scrapyards that do allow sales, you can find items at a third or less of their retail cost: rolls of welded-wire fencing, chain-link gates, chain-link fencing, rolls of hardware cloth, roofing metal, latches, hinges, bolts, screws, hand tools, and all kinds of brass, aluminum, or steel tchotchkes and gewgaws for decorating your habitat.

Yard Sales

Our best luck has been at yard sales in the country or in middle-class neighborhoods. Prices are mostly negotiable, and many objects can be had for a tenth of their retail value or less. Be on the lookout for hand tools, power tools, power cords, plywood, planks, and so on.

Thrift Shops, Junk-tique Stores, and Flea Markets

A good thrift shop has castoffs in good condition. What we call a "junk-tique" store is a thrift shop with higher aspirations: it may have some actual antiques, and the prices will be higher. Flea markets can run the gamut from sad to sensational. Generally, prices at all three places are negotiable, especially if you're interested in more than one object: "If I buy items A, B, and C, will you take x dollars?"

Keep your eye out for vintage chicken feeders, waterers, egg baskets, hand tools, vintage or reproduction chicken signs, books on raising poultry, and so forth.

Residential Construction Sites

If you see someone working on new construction or a renovation, you could ask for permission to scope out their piles of scrap before they haul them off. Sometimes there's little to find, but you could also hit a jackpot of plywood, planks, rolls of fencing, metal flashing, or unused roofing shingles.

Curbsides

Some people will throw lightly used things in the trash, but others will put such objects out at the curb for others to scavenge. In these situations, I've snared rolls of fence wire, lumber, and metal dog crates (to quarantine broody hens).

We rescued Miss Piggy from a junk-tique shop for a few dollars. She watches over the chickens from atop the pen.

10 Coop Design

I'll bet one of the first things that crossed your mind after deciding you wanted chickens was "What about a coop?" (And by "coop" you're probably also thinking of what I call "hen habitat," which includes the fenced run attached to the coop and all the other features that go with chicken-keeping: a waterer, a feeder, feed storage, etc.) So, yes, what about the coop? What is it really doing for you? What does it need to accomplish? What are the cost-effective and low-maintenance ways to accomplish those goals?

Why Even Have a Coop?

People in many countries around the world keep chickens without a coop or run. But for the urban hipster, the suburban homesteader, the rural hobby farmer, and everyone in between, going coopless means at least four unsatisfying outcomes:

1. Without a coop's nest box, hens will lay eggs nearly anywhere. For hens, dark, secluded corners and crannies look like the beginnings of a nest. For you, that means a time-consuming, daily Easter egg hunt. Cheer up! Any eggs you miss won't go to waste — they'll feed your local snakes and vermin, boosting their population. You'll be missing out on some eggy breakfasts and frittatas, but you'll be supporting the local wildlife.

2. Without some containment, not only will your foraging and scratching hens damage or even destroy your gardens and lawn, but their habit of continuous pooping will organically decorate your walkways, patio, porch, and deck.

3. During the day, your foraging hens will make tempting targets for raptors like hawks and eagles.

4. At night, the hens will follow their ancestral calling to roost on some low-hanging branches of your trees or shrubs. That will get them high enough to be safe from nocturnal and earthbound stray dogs, but they won't be safe from creatures that can climb, such as stray housecats, native gray foxes, raccoons, weasels, minks, and possums.

See what I mean about the limits of chickens' domestication? That's where we henkeepers come in, with our coops.

Construction Overview

Before we get into the nitty-gritty of building your coop, I want to share some big-picture thoughts about coop building.

Get Floored

The first step is to raise the coop up off the ground. Any raw wood touching the soil will rot away. Get the wood at least 1 foot above ground so that splashing rain can't keep it damp enough to rot; you only want to build your coop once.

The cheapest foundation for a coop is a pair of cinder blocks stacked to support each corner of the wooden pallets (see page 174). Cinder blocks are 8 inches high, so a pair raises the pallets 16 inches above the ground. This height also essentially creates a moat (see below).

Given the small size and weight of a backyard coop, it's not necessary to dig footings below the frost line. The movement of the soil during freeze-thaw and drought-monsoon cycles won't damage the coop. The worst that could happen is that a coop door may become sticky. It's easier to adjust a door with a jigsaw or drill than to dig a foundation.

Every Castle Needs a Moat

Lots of critters would love to snuggle up with your hens and their eggs — just once, of course. And it's not like they have jobs to go to. They've got all night to work their way into a coop. There's no point making it easy for them. Since your flock is counting on you to keep them safe, you'll stuff fence wire scraps into all the gaps in the coop walls and put latches on the hatches. You'll make your coop into a regular fortress to defend your fair ladies from pirates, bandits, and Vandals and Visigoths. But will your castle have a moat?

Our coop, like many others, is raised up off the ground on posts. Without being able to stand on the ground, rats will have a harder time gnawing their way through the wood or wriggling around a door, and raccoons will find it tougher to pry off a loose board or pick your combination lock.

Beyond deterring vermin, I find there are four other advantages to building our coop perched above a "moat":

- **Cost.** Suspending the coop floor 16 to 32 inches above the ground means we don't have to buy expensive rot-resistant wood for the floor. Of course, the four posts in the ground should be cinder blocks or a rot-resistant wood such as cedar, redwood, or pressure-treated wood. The floor joists and the floor deck can be any old pine, spruce, fir, or even whole pallets.

- **Cover.** We left our pen open to the sky so the rain would help the chicken poop and mulch decompose and draw earthworms and other small edible creatures into the pen. When it rains our queens of the castle can choose between a refreshing rinse or staying dry under the coop.

- **Dust bath space.** After many days of rain, the ground in an open pen will be too wet for a good rejuvenating dust bath. But the area under the coop will be dry (and more private for bathing!). That's why I don't put mulch under the coop — so the bare, dry dirt will be accessible.

- **Food preservation.** Our gravity-fed feeder has an open top, so we keep it under the coop. This means it stays dry during rains. Just as important, there isn't enough room for the hens to roost on it and "fowl up" their food supply.

- **Back preservation.** The henkeeper should be the primary client and the hens should be secondary (although it often doesn't feel that way). When the time comes to clean out the coop, the ceiling should be at a comfortable working height for the primary client. That way your ladies-in-waiting won't have to wait too long for the servant to clean up their chambers.

Construction Diary: Staging a Coop

Here's how I helped a young first-grade teacher named Katie Ford and her stepfather, Johnny Ford, build a bigger coop so she could keep more birds on a limited budget. She wanted to keep two types of chickens in the coop for breeding — Lavender Orpingtons and Copper Marans — so she would need to keep them separate. I joked that the coop would be like a duplex and christened the project Katie's Kooplex.

The next day Katie and Johnny gathered about two dozen pallets. I picked up some lumber and cinder blocks, and we got started.

HOW WE DID IT

1. I laid out a footprint on the ground based on the dimensions of the pallets. Six pallets gave us a floor of 12 feet wide by 7 feet long, or 84 square feet. Chickens require about 3 square feet apiece in a coop, so this would be more than enough room for the two dozen chickens Katie envisioned. Most backyard coops only need to be big enough for 5 to 10 hens.

2. We placed cinder blocks to support every corner of the six pallets that would make up the floor framing. Every floor pallet was attached to every other floor pallet it touched with a trio of 3" exterior screws (one in each end and one in the middle). We covered the pallets with two and a half sheets of ¼" plywood, screwed to the pallets with 2" screws, and cut off any excess plywood. That made for a smooth floor that would be easy to clean out.

3. We stood some more pallets on their sides to form 40"-tall walls, turning the pallets so that their 2" thick planks would contact the floor and the roof. We screwed them in place with a trio of 3" screws at every point of contact. The 1" thick planks of the pallets, now vertical, would function as studs for support and for attaching exterior siding.

4. Using a jigsaw, we cut openings in the wall pallets for the nest box and for chickens and people doorways.

5. We framed a ⁶⁄₁₂-pitch (45 degree angle) gabled roof using new lumber and screws, covering it with salvaged metal roofing panels of different colors.

6. To cover the gaps in the walls, we screwed on planks liberated from pallets by cutting their nails with a reciprocating saw and metal-cutting blade (see page 162).

With some hardware cloth covering the ventilated gables (see page 116), nest box and doors installed, feeder and waterer in place, and a walk-up plank attached, Katie's Kooplex was ready for business.

Building a Coop from Free Pallets

There's no need to spend lots of money on materials for a coop. Many tens of thousands of wooden pallets are sent to the landfill each year. With a drill, screws, and jigsaw you can put up the framing for a coop in a day or two.

Tools and Materials

A backyard coop for five to eight hens requires only about fourteen pallets: two for the floor, five for the wall framing, three for the exterior siding of the walls and four for the roof framing.

Most experts recommend 2 to 3 square feet of floor space per hen. A floor composed of two standard pallets (see next pages) makes enough space for five to eight hens. If you want to house more hens, lengthening the coop with a third floor pallet (and a commensurate increase in other components) would allow you to add another three to four birds.

For best results, use pallets that are all of the same dimensions. The most common size is 5 inches thick, 40 inches wide, and 44 inches long. A few wall pallets will have to be cut to fit the length of the floor. Where the pallets meet each other, you'll have to drive the 3" screws in some creative diagonal angles, but that will work fine.

Do Screw It Up

Nails are cheaper than screws, but not everyone has the upper body strength or the sure eye needed to drive enough nails to hold a coop together without smashing a thumb or wishing the coop would build itself. Also, banging away with a hammer can knock the pallets out of alignment.

I recommend fastening pallets with 3"-long exterior decking screws and a corded drill. Whether using nails or screws, you need more than half of the fastener's length to go into the second piece of wood to have a secure connection. Decking screws will go through pallet wood without needing a predrilled hole to get them started as long as the pallet is made of lightweight pine. Very rarely you may come across pallets made of dense oak, but they weigh three times as much, so it will be pretty obvious they aren't pine.

HOW TO MAKE IT

A PALLET FLOOR

WHAT YOU NEED

- 2 pallets (5" × 40" × 44")
- Stick or marking paint
- 12 cinder blocks (8" × 8" × 16")
- Level
- Wooden shims (optional)
- Jigsaw
- Corded drill
- 3" exterior decking screws (see page 173)
- One ¼" × 4' × 8' sheet of plywood
- 2" screws

Construction Tip: In your chosen coop location, lay out two pallets to mock up the coop floor on the ground (allow 2 to 3 square feet per bird). The sides that touch at the center of the coop should be the 44" sides, so that a 2×4 side of one pallet is touching the 2×4 side of the other pallet. Mark the locations of the corners of each pallet on the ground with a stick or marking paint.

WHAT TO DO

Set the cinder blocks.

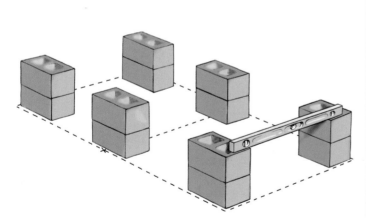

1. Remove the pallets and stack the cinder block pairs about 1" inside the corner marks so the pallets will hang over the blocks by an inch. This will keep rainwater from seeping in between the cinder blocks and pallets and causing the pallets to rot.

2. Use a level and a straight plank to check that the tops of the cinder blocks are all at the same height.

Place the pallets on the cinder blocks and screw them together.

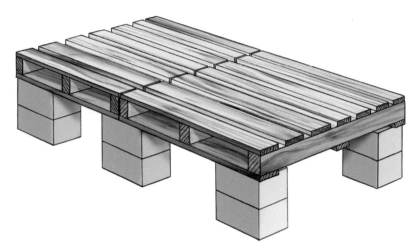

3. Using the drill, put a trio of 3" screws through each edge of a floor pallet where it meets its neighboring pallet: one screw at each end and one in the middle. Double-check to make sure every edge is secured before moving on. (There's no need to fasten the pallets to the cinder blocks. Short of a hurricane or an earthquake, the coop's weight will keep it in place.)

Lay a sheet of plywood for a floor.

4. Hold the plywood sheet in place with a couple of clamps.

5. A 2" screw or nail every foot or so will be long enough to hold the plywood down securely. That smooth surface will make the floor much easier to clean out.

6. With a pencil and level, mark where the plywood hangs past the edge of the pallets.

7. With a jigsaw, carefully cut off excess plywood without cutting into the pallets.

Wall It In

Your coop walls don't have to be tall enough for a person to enter the coop often if you're using an exterior nest box. A single row of pallets and screwed to the floor makes a wall tall enough for your hens.

Project Notes

- Before you begin, examine your pallets closely. Each one has two long sides defined by solid 2×4 planks (with a third one in the center) and two short sides that are open. Turn the pallet so that the long, solid sides form the top and bottom of the walls. That way you can drive 3" screws through the bottom 2×4s, then through the plywood and into the pallet floor.

- When you build the roof, you can screw pallets or rafters to the 2×4 on the top of the pallet wall.

- With the framing up, use a jigsaw to cut out openings for chicken doors, cleanout doors, and exterior nest boxes.

- Temporarily leave one wall pallet out, so that you can get inside the coop to frame the roof before installing the siding. Then, with the money you saved on lumber, go shopping for chicks.

Core Tools

Here are the basic tools you will need for the remaining building steps. Most steps require a few unique items of their own; those are listed at the start of each step.

- Corded drill and assorted drill and screwdriver bits
- Box of 3" exterior-grade decking screws
- 2" screws
- Jigsaw
- Power cord
- 4' level
- Speed Square
- Tape measure
- Marking pencil
- 4 quick clamps

HOW TO MAKE IT

PALLET WALLS

WHAT YOU NEED

- Core tools (see page 177)
- Five standard pallets (5" × 40" × 44")

WHAT TO DO

Attach the end walls to the floor.

1. Stand a pallet at one end of the coop with the 2×4 side down. The wall pallet's width should match the floor's, and the planks of the wall pallet should be flush with the floor edge. Use three 3" screws to secure the wall pallet to the floor, one screw at each end and one through the middle of the 2×4. Do the same at the other end of the coop.

Attach side walls to the floor and one end wall.

2. Lay another pallet on its 2×4 side so that its planks are flush with the floor edge. The edges of each wall pallet should meet nicely and both walls should be plumb or very close to it. Use three more 3" screws to secure the side wall pallet to the end wall pallet, and another three 3" screws to attach the side wall to the floor. Repeat this step on the other side of the coop.

Prepare two narrow wall sections, and install one.

3. On both sides you'll have a gap remaining that is narrower than a pallet. Measure the distance carefully on each side and cut two pieces (possibly from one pallet) to fit in those gaps. Make sure that each piece has a 2×4 side that will rest on the floor. Install one of these pieces with three screws on each side that meets another wall or the floor. Set the other cut-down piece aside so that you can easily get inside the coop for the next steps, like roofing and screening as shown on pages 182–187.

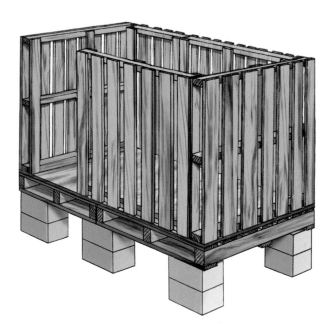

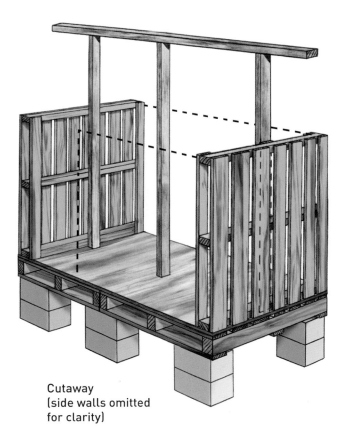

Cutaway
(side walls omitted
for clarity)

Cut ridge poles for roof.

4. If the pallet walls are the standard 40" high, cut three 2×4s (called "studs" at the lumberyard or big-box store) so they are 56" long (or 16" higher than the walls if your pallets are a different height). These are the ridge poles and will support a 2" × 4" × 8' stud, the ridge beam that holds up the center of the roof.

Install ridge poles to support the ridge beam.

5. Find and mark the center point of the top of the two end walls. Place one pole against an end wall, with one of the pole's wide faces against the wall and centered on the mark. Use the level to check the pole for plumb. Use three 3" screws to secure the pole to the three 2×4's in the end wall. Repeat on the other end wall.

Install a ridge beam and a center pole to support the roof pallets.

6. Lay the 2" × 4" × 8' ridge beam on top of these two poles with the wide side against the tops of the poles. Make sure that 9" of the beam extends past each of the poles. Drive two 3" screws through the ridge beam and into the top of each ridge pole. Double-check that each end extends past the pole by 9".

7. Find and mark the midpoint between the two poles on both the floor and the underside of the beam. Line up the third pole on these two center points and make sure it's plumb.

The pole doesn't have to line up precisely on the center points, but it does need to be plumb to keep the beam from sagging in the middle. Drive two 3" screws through the beam and into the top of this middle pole. Drive two 3" screws diagonally through the base of the pole and into the floor.

Dry-fit the remaining section of wall.

8. Insert the last narrow piece of wall without screwing it into place — that's called "dry-fitting" — to make sure everything fits together and is sturdy. Don't attach this wall piece permanently until you install the roof pallets and screens.

Add a Roosting Bar

This is a good time to install a 2×4 stud cut to 76"–78" to serve as a roosting bar. Lay each end of it on the middle 2×4's of each end wall (you may have to use the jigsaw to cut out part of the pallet to make room for the roosting bar). The wide side of the bar will be resting on the wide side of the end wall 2×4's. The bar will run parallel to the beam and touch the three poles. Put one 3" screw through each end of the roosting bar into the end wall 2×4. Put one 3" screw diagonally through the bar and into the middle pole to keep the bar from sagging.

Put a Lid on It

With the roof supports and roost bar in place you can install the four roof pallets, the plywood roof, and waterproof roofing. A helper, a ladder and some patience will be very valuable here. Working above ground can be challenging and the pallets will be resting at an angle, so go slow and have help.

HOW TO MAKE IT

A WATERPROOF ROOF FOR THE GIRLS

WHAT YOU NEED

- Core tools (see page 177)
- A strong person on the ground outside the coop and another person inside the coop with the power drill and screws
- 6' stepladder
- Four pallets
- Drill, bits
- Two 4' × 8' sheets of ¼" plywood (not pressure-treated)
- Four quick clamps
- 3' wide × 33' long roll of mineral surface roofing (you will use at least 24' of roofing)

- Retractable razor knife
- Hammer
- 1-pound box of 1" roofing nails
- 10' length of galvanized ridge cap
- Two dozen 2" metal roofing screws
- Matching roofing screw bit
- Metal snips
- Automatic door (or one 8" × 12" piece of ¾" plywood, and two hinges)

- Three plywood sheets ¾" thick and up to 2' × 2' (not pressure-treated) for the two cleanout doors and the chicken-keeper door (doors can be purchased in those dimensions or cut to size from a full 4' × 8' sheet at a big box store).
- Best Nest Box (see chapter 7)
- 6 hinges and matching screws

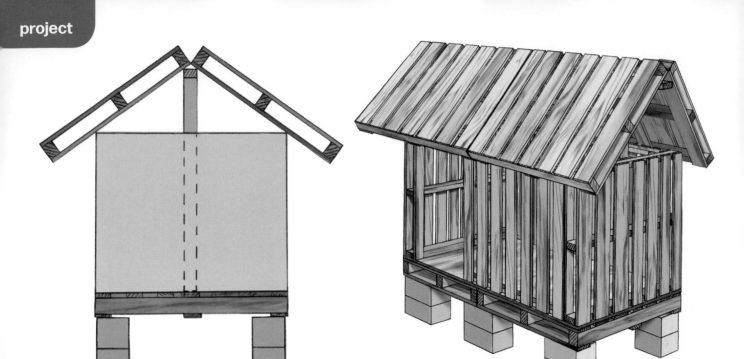

WHAT TO DO

Install a pallet roof on top of the framing.

1. Find the midpoint of the length of the ridge beam (it should be right over the middle ridge pole) and mark it. Also mark the center of the *width* of the beam at that point. Find the midpoint of the length of the top of each side wall and mark it.

2. With one person standing outside the coop and holding the low side of a roof pallet, place the pallet so that an end with a closed edge (with a 2 × 4) rests on the ridge beam and the rest of the pallet rests on a side wall.

3. The person inside the coop should position that first roof pallet so that one open edge lines up with the midpoints you marked on the beam and the sidewall. The other open edge of the roof pallet should meet the end of the ridge beam and extend 4" beyond the end wall.

4. Make sure the closed edge lines up with the centerline of the beam, so there will be room

for the roof pallet on the other side to also rest on the beam, with the roof pallets just touching.

5. Double-check that the first roof pallet is lined up with all its marks. The roof pallet should extend beyond the end wall by about 4" and beyond the sidewall by about 8". Most important is that it lines up on the center marks and the high side rests on the ridge beam, so the other three pallets will fit together nicely. Clamp the pallet into place.

6. Drive three 3" screws up through the ridge beam into the roof pallet. Double-check that it's still lined up on the three center marks; move it and reattach it if necessary.

7. From inside the coop, drive three 3" screws up through the 2 × 4s of the side wall and into the roof pallet to secure it.

8. Install a second roof pallet so it touches the first roof pallet on top of the beam and attach it as in steps 2 through 7.

9. Using the same process install a third and fourth roof pallet.

Install plywood over the pallet roof.

10. Lay one plywood sheet on one side of the roof so that each end (on your right and left) hangs over the pallets by the same distance. Clamp the plywood to the roof pallets. Do the same with the other sheet of plywood.

11. Adjust the two sheets of plywood so that they just touch over the centerline of the ridge beam and re-clamp them. If your pallet dimensions are different, make sure that the plywood either meets the bottom and sides of the pallets or can be cut to just 2" beyond the edge of the pallets.

12. Screw the right and left ends of the plywood to the roof pallets with 2" screws, spaced 8" to 12" apart. Remove the clamps.

13. If the plywood extends beyond the pallets, use the level and pencil to mark on the plywood the location of the sides, lower edge, and upper edge of the roof pallets. With the jigsaw, following your marks, trim off anything more than 2" beyond the pallet bottom and sides (but not the top). Screw in the sides and lower edges every 8" to 12".

14. Below the line you made near the top of each sheet of plywood, drive screws every 8" to 12" through the plywood and into the roof pallets.

15. Put in another dozen 2" screws scattered evenly across each sheet of plywood to pin it down to the pallets. **Note:** At the very top where the two sheets of plywood meet there is nothing to screw them to, but they will be fine for now.

16. Install the final short piece of the sidewall.

With the plywood in place for the roof, you're ready to "dry in" the coop by making the roof watertight.

Install waterproof roofing atop the plywood sheet.

17. Unroll about 10' of the 3'-wide roofing material on the ground. Use the marker, tape measure, Speed Square, level, and retractable razor to mark and cut three lengths of roofing that are 2" longer than the plywood so it will overhang by 1" on each side. Roll two up and set aside.

18. Roll out one section horizontally onto one side of the plywood roof. Adjust the roofing material so that it extends 1" beyond the bottom edge and both sides; this way the rainwater will flow off without wetting the wooden part of the roof. Clamp into place.

19. Nail the roofing material in place with roofing nails spaced every 8" to 12" along the sides and bottom and into the edges of the roof pallets. Don't put nails along the top edge of the rolled roofing. Spare your fingers by holding the nails with a pair of pliers.

20. Repeat steps 18 and 19 on the other side of the roof.

21. Use the third piece of roofing material to cover the peak of the roof. Center the piece over the line where the sheets of plywood meet, and gently bend it over the ridge so it extends as evenly as possible down each side of the roof. The center piece of rolled roofing should overlap the lower strips of rolled roofing by about a foot or so depending on the dimensions of your pallets. The left and right sides should extend 1" beyond the plywood to match

the other strips of rolled roofing. Nail the center piece of roofing into the pallets along both sides and the bottom edge.

22. Bending the center piece of roofing over the roof peak can crimp and crack the material and lead to leaks over time, so lay the 10' of galvanized ridge cap over the peak and center it by eye. Bring one end even with the edge of the roofing material.

23. Drive the 2" metal roofing screws along both edges of the ridge cap about 1" from the edge and 8"–12" apart.

24. Using metal snips, trim off the end of the ridge cap that is extending beyond the roof.

Congratulations! You now have a watertight roof over your coop!

Construction Tip: Door Dimensions

The dimensions of the cleanout and keeper doors will vary depending on the particular pallets and the spacing of their slats. You don't want to cut through the 2×4s that run through the middle of the wall pallets (you'll need them for attaching the additional siding) or the ones along the floor (you'll need them to keep the doors from opening inward). These three doors will likely be no more than about 16" tall. They can still be up to 24" wide but no more — wider would mean you would have to cut through the vertical plank on the inside of a wall pallet, and that plank helps it keep its shape and strength.

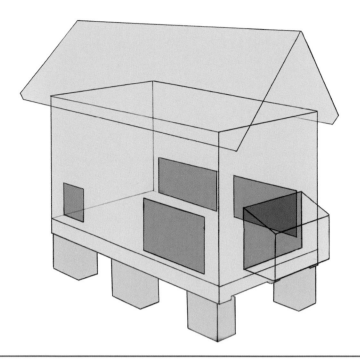

INS AND OUTS: DOORWAYS

Cut openings for doors and nest box.

1. Determine which walls will have which openings. The opening for the automatic door will be in the wall that will be inside the fenced pen. The chicken-keeper access door can be inside or outside the pen. The two cleanout doors and the nest box should be outside the pen. The cleanout doors should be in a wall with room outside for installing a couple of compost bins, or with easy access for a wheelbarrow.

2. Using tape, level (for marking straight lines), Speed Square (for marking perpendicular lines), and pencil, mark all five openings. The dimensions of the automatic door opening will depend on the brand you buy. If you forgo an automatic door, cut that opening to roughly 8" wide by 12" tall.

3. Mark the opening for the Best Nest Box to be 24" wide and 12" high. Measure and position the opening so that the flange of the nest box floor will rest on the horizontal 2×4 at the bottom of the wall pallet.

4. With the five openings marked, cut them out with a jigsaw.

5. Transfer the dimensions of your door openings to the piece of ¾" plywood and cut the doors out with the jigsaw. Dry-fit the pieces to double-check your work.

6. While a helper holds the doors in place, position the hinges and with the pencil mark the spots on the doors for the screws. Predrill holes with a bit that matches the diameter of the shafts of the screws. Attach the hinges to the doors.

7. Have your helper hold each door in place on the coop and mark where the hinge openings fall on the wall. Predrill and then attach, as above.

8. Attach a latch (as discussed on page 92) to secure each door.

9. Attach the nest box as described on page 138.

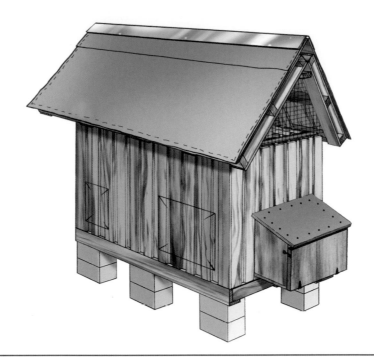

COVER ALL GAPS TO KEEP YOUR GIRLS SAFE

With the doors installed it's time to button up the walls of the coop. Scrap lumber can be cut to fit for this purpose, but if you have just three more pallets, their slats can be removed and will fit perfectly over the openings in your coop. Use fence wire in gables for ventilation. A helper will make this project go faster.

WHAT YOU NEED

- Core tools (see page 177)
- Three pallets
- Something to lean the pallets against, like a tree or a post

- Sawzall with metal-cutting blades
- Pair of sawhorses or a table, to stack the slats on
- Metal snips

- Chicken wire, fence wire, or hardware cloth
- ¾" fence staples
- Pliers
- Hammer

WHAT TO DO

1. One at a time, lean the three pallets against a post, building, or sturdy tree. With the Sawzall and metal-cutting blades, take the pallets apart, as described on page 162. When all three are dismantled, you should have enough slats to cover all the gaps in the walls of the coop. Stack the slats near the coop.

2. Have a helper hold the first slat vertically over a gap in the wall. Mark the three places on the slat where it aligns with the three 2×4s of the wall pallet.

3. Neatly stack about six slats on the sawhorses, with the marked slat on top. Using a drill bit slightly greater than the diameter of the threads of the 3" screws, drill down through the three marks as deep through the stack of slats as the bit will go. Using one of these predrilled slats on top of another neat stack, repeat this process until all the slats have been predrilled.

4. Hold a slat evenly over a gap in the wall (or it may fit just inside the gap) and drive three screws through the predrilled holes and into the 2×4s. Continue around the coop.

5. Cut some slats to match the wall height above the doors and nest box and install as described above.

6. To cover the triangular gable openings, cut fence wire to extend 2" on all sides of each opening. Hold the wire in place on the inside of the gables (a helper can do this from the outside the coop by hand or with some pliers) and nail the excess wire into place from inside the coop.

7. To fill the gaps in the open ends of the four roof pallets, use the metal snips to cut the fence wire to dimensions that roughly match the openings. While a helper holds the wire in place, hold staples with a set of pliers and drive into place with the hammer, Set staples about 4" to 8" apart. Alternatively, the open ends of the four roof pallets can be covered by slats or scraps of wood.

8. Final step: Think like a predator and look for any small gaps into the coop that a snake or rat may be able to get through. If you find any, roll up some chicken wire into a ball or a roll that can be pushed into the gaps as shown in the photo on page 116. Drive one or more screws through the wad of wire and into some wood to pin it in place.

HENKEEPERS I'VE KNOWN

Author Amy Stewart's First-Lady Chickens

Amy Stewart raises hens for entertainment, affection, and eggs. She also paints their portraits, which she sells online and at a local gallery. Amy is better known for her best-selling nonfiction books *The Drunken Botanist, Wicked Bugs, Wicked Plants, Flower Confidential, The Earth Moved,* and *From the Ground Up,* and, most recently, for her novel, *Girl Waits with Gun.* She answered my questions about her birds (in two dimensions as well as three) over the course of a sunny afternoon spent in her ⅙-acre backyard garden and upstairs in her painter's garret in a historic bungalow in downtown Eureka, California.

FH: What does it take to paint a chicken's portrait from a photo?

AS: The trick is holding the camera and tossing some chicken scratch and getting them to look up all of a sudden. You have to have your camera set on that sports setting so you can get lots of shots real fast. You have to draw them out into the light. But they want to mill about in the shade.

I like to get them head-on with both eyes, but they're not really looking at you when they're looking straight ahead. Their eyes are on either side of their heads. The pose they want to be in is with their butts in the air and their heads on the ground, so I have to take hundreds of pictures to get one good one. To get them doing that in nice light is an even bigger struggle.

I have spent endless hours out there photographing them only to get three good images that I can use. It's kind of a two-person job. If I can I get [my husband] Scott out there, he'll be keeping them in position and trying to get them to keep their heads up with some scratch.

FH: Are any of the chickens more cooperative than the others?

AS: They're all equally uncooperative. It's not part of their normal thing that they do. It's nice to get them isolated, up on a table or on a chair. Of course you get one chicken up there and the other chickens want to know what's going on, so they jump up and knock that chicken off and battles ensue, and then they're focused on each other. It's complicated. That's the struggle with chicken painting.

FH: Is there an easier way?

AS: It's kind of sad to say, but the easiest place to get good pictures of chickens to paint is at a poultry show. They're in a cage and they tend to just stand there and look because there's not that much to do.

FH: Do your chickens' personalities show at all in their facial expressions?

AS: I think so, because I know them. I've painted other peoples' chickens, too — they all have their little personalities. That's the problem with chickens: their face is fixed. The reason we love dogs so much is that they can wrinkle their brows and they can smile — things that read as human. But chickens have this little permanent frown, which is not their fault, and their beaks curve down. They can't have expression with their eyes because there are no features around their eyes.

Chicken portraiture is tricky in that way, but I really like getting them photographed like people, where they're looking at the camera and it's

> **The pose they want to be in is with their butts in the air and their heads on the ground.**

head and shoulders. My idea was to make the paintings look like serious portraits, like family pictures. You know, you go up the stairs to a portrait studio and you get your picture taken. I wanted them to be like that. Like head shots. You try to get their personalities a little bit that way.

FH: What kind of personalities do your birds have?

AS: I wanted visual variety [with my chickens], but you realize with any pet it's all about personality — what are they like to live with. How they look is secondary.

Abigail [a Golden Laced Wyandotte] is the shy one, very suspicious. She does look a little suspicious in that painting. She's the one, you bring out food and she looks like, "Well you're trying to poison us, I know you are. I know that today is the day you're going to kill us." She's very conspiracy-minded.

Lady Bird [a Buff Orpington] is the sweetest one. I think she looks like she wants a hug in that picture. She follows you around like a little puppy dog and wants to be picked up and be your friend. If she were a cat she'd be the one rubbing up against your legs everywhere you go. Dolley was great, was my favorite chicken. It was so sad when she died. Really smart. Top of the pecking order. First one to figure something out. An Ameraucana. Smart crafty little birds, more birdlike in every way.

FH: How long have you had chickens? And how many do you have?

AS: Eight years. Abigail is the only one from the original batch; the other three have all died. Tumors. We have a bird vet we take them to — I'm not about to kill them with my bare hands. What I think is happening is that hatcheries are not breeding for longevity, of course. Like dogs . . . some breeds, their hips go. It's not like a parrot getting sick, as pets go. It's really a different thing dealing with long-term health issues [of chickens], because that's not what farming is about.

In addition to Abigail (Adams), there's now Lady Bird (Johnson) and Ida (McKinley, an Ameraucana). They're all named after First Ladies. No longer with us — Eleanor [a Rhode Island Red] died at five years, Dolley [an Ameraucana] at six, Bess [also an Ameraucana] at seven. You're gonna get 15 years out of a cat. Out of six chickens, three have died. Three animal funerals in five years. That's too many.

They've also had many gynecological issues. They will get a prolapsed uterus, which involves latex gloves and poking it back in there. It's happened when I'm gone and my poor husband is here. Scott said, "This is not a job for a man," as I was guiding him on the phone. I felt so sorry for him.

> **My idea was to make the paintings look like serious portraits, like family pictures.**

FH: Do your chickens free-range here?

AS: They free-range all day unless we're away. The plants back here are survival of the fittest. It's what the chickens don't destroy. There used to be a lot of self-sowers, but they're all gone since they eat the seeds. The chickens eat the annual weeds, too. Just Himalayan blackberry, bindweed, and dock are [weed] problems. As you can see, nothing gets cut down around here. It's a forest.

The nice thing about this garden is that the chickens love it. This thicket of plants — they can get up under there. They have probably laid eggs up under there. It's nice for chickens to have shrubs and cover like this. There are plenty of bugs for them to find. I have a ton of raspberries — no idea what variety. [To one chicken:] Ida, want to jump for it? Lady Bird would jump for it. It's a berry. You know you want it. You're a good little jumper.

Got apple trees here: Honey Crisp, Liberty. Only problem with the apple trees, chickens jump up in the tree and take a bite out of every apple to see if it's ripe.

FH: Do you clip their wings?

AS: No. Before we built the higher fence, they did fly over to the neighbors once or twice by accident. Chickens are homebodies. They don't want to go anywhere, so I'm not at all worried about them flying away. [To the chickens:] We call you "chicken" for a reason.

FH: What's their coop like?

AS: This was just a shed that was here when we bought the house. I just got this [a chicken waterer that screws onto a water cooler] — pretty excited about it — so the water stays clean. With my husband and my dad, we added the run with hardware cloth on top and a cement floor. But mice can get in. We have a mice problem. Door is rotting. Neither of us is handy.

FH: How did you get started painting?

AS: I started painting when I moved up here [to Eureka from Santa Cruz] 10 years ago. I wanted to take a little drawing class so I could sketch the garden, but I couldn't find a good ongoing class I could take. An oil painting class, by a painter I really liked, Linda Mitchell, fit my schedule.

Turns out I really liked oil painting. I thought it was going to be too hard, too much work, a real hassle. But oil painting is a lot like writing. It's about revisions and editing. You rework it, over and over. All oil painting is repainting. You do one on top of the other. As a writer, I know how that goes. My teacher would say, "Wipe that whole thing off and start over," and I'd say, "Okay." That doesn't bother me. I really liked it for all those reasons.

I paint small because I don't have a lot of time. I can finish a small painting — 5 by 7 inches — at one go.

[We move up to her home's attic. It's a painter's garret with an easel, a small woodstove, love seat, and low bookshelves. The knee walls and exposed roof rafters are painted vanilla white. North-facing windows feature a view of Eureka harbor. A 15-year-old cat named Kitty follows us.]

FH: Wow.

AS: It's all old-growth redwood cut 2 inches thick. [The floors, walls, and rafters were cut from 2,000-year-old trees and built by the carpenter who also worked on the biggest mansion in town.]

I paint on black gesso, so the background starts off black. Some of these paints are $25 a tube. They've gone up because of the cost of minerals. An expensive hobby, but it's nice to do something that's not for money or career or in front of a computer. I don't have to do it any particular way. I don't have to be good at any particular thing. My friends who are artists have to do all this career crap. I don't need to win awards or add to my résumé or any of that crap.

I sell them, but just to get rid of them. They pile up. I've probably sold 100 to 115. It depends on how much I'm home and how much I'm doing. I just need them to go somewhere. I don't want to look at my own art on the wall. My painting teacher has some gallery space so some are over there, but the rest of them I sell on the Internet. The chicken paintings . . . people actually love them and they're fun to do.

FH: What's next?

AS: Our latest challenge is figuring out an easier way for our elderly and flight-challenged birds to get down from the high roost they've chosen for themselves. Up is easy, but gravity makes down particularly hard on their weird old feet. A ladder? Straw bales? Ah, the challenges of chicken-keeping.

(Learn more about Amy Stewart at http://www.amystewart.com/ and http://www.amystewart.com/paintings/.)

In her sunlit garret, Amy paints chicken portraits from photographs.

An oil portrait of one of Amy's favorite and much-missed hens: an Ameraucana named Dolley Madison.

Amy Stewart's hens — Lady Bird Johnson in front and Abigail Adams — free-range in a garden of plants that she describes as "survivors."

Egg-cellent Questions

Q: Do you need a rooster to have eggs?

A: No, your hens don't need roosters to help them produce eggs. Female animals of all species produce eggs. The real question is whether the eggs have been fertilized or not. Fortunately for us, female chickens will lay eggs whether or not they are fertilized. Without a rooster, the eggs you harvest will be unfertilized, but they will taste and look exactly the same as fertile eggs.

Q: Why, then, do people have roosters?

A: Good question. A really good question! Roosters are noisy. They may crow loudly at any time of day. That's why many urban chicken ordinances outlaw roosters. They can also be very aggressive. Most people who dislike chickens have a childhood story about being attacked by roosters.

If you want to have fertile eggs, however, so that you can hatch your own chicks, you will need a rooster to get that job done. A hen that has laid fertile eggs will tend them, and they will hatch without much attention on your part. And the chicks will be accepted as part of the flock.

Finally, roosters love their hens and will protect their harem from predators — that is, unless the predator is big enough to kill or chase off the rooster.

Q: Do eggs of different colors have different flavors or different health benefits?

A: Nope. Eggs of the same size, regardless of color, have the same calories and health benefits. Eggs of any color will taste just the same.

Q: Where do the different egg colors come from?

A: The genetic heritage of the hen determines the color of the eggshell. The shades may vary somewhat according to the hen's diet that week. The color is laid down as each egg's shell is developing inside the hen.

The one exception to that is the very dark brown color of eggs from Marans. That dark coating is a superficial layer that is applied as the egg moves through the oviduct. The longer it takes to move through the oviduct, the darker the egg will be.

Q: How many eggs will a chicken lay?

A: I've read that a hen can lay one egg as frequently as every 25 hours. An Australorp set a world record of 364 eggs in a year. But most hens will lay three to five eggs per week, tops.

Hens are at their most productive in their second year of life. Their productivity is at a plateau for the next year or two, and then slowly their productivity slows down for three or four years until they are producing only one egg a week, and then none.

Q: What are the advantages of backyard eggs?

A: All foods – eggs, produce, meats, and so on – decline in quality from the instant they are harvested. You will never buy an egg that is as fresh, nutritious, and delicious as one harvested from your backyard mere hours after it has been laid. If your hens are eating bugs, weeds, and kitchen scraps, then the whites will be thicker, the yolks darker, and the combination more appetizing than with store-bought eggs.

Chickens getting a varied diet may lay eggs that contain half the cholesterol of store-bought eggs. Eggs from backyard birds may also have twice as much or more of vitamin E, beta-carotene, and omega-3 fatty acids as factory eggs.

Q: Do I need to harvest the eggs right after they are laid and refrigerate them?

A: No. Freshly laid eggs are covered with a coating from the oviduct called a "bloom," which keeps the eggs fresh for up to a month without refrigeration. If you are on vacation the eggs can wait in the nest box until your return. They will still be fresh and safe to eat.

Q: Some eggs are dirty. How should I wash them?

A: Wipe off the dirt, which is most likely poop, with a DRY paper towel and store the eggs on the countertop. You don't have to wash them until you are ready to cook them. There is a small chance that the feces may have salmonella (probably not, but you may want to play it safe). Wash that egg, and your hands, before cracking it.

If you wash the egg as soon as you harvest it, then you'll also be washing off the bloom that keeps a room-temperature egg fresh. Washed eggs must be stored in the fridge.

Q: Will hens ever eat their own eggs?

A: It's rare for a hen to intentionally break an egg to eat it. But if an egg unintentionally breaks in the nest box, almost any hen will sample and eat the yolk, whites, and even the shell. This is called egg cannibalism, and it's a habit you don't want to see get started. If it does start, you want to break it right away by harvesting eggs as quickly as possible.

You may also need to isolate the offending hen in a separate pen, such as one made from a large dog crate. Even if you can't stop her from eating her own eggs, the other hens' eggs aren't at risk and hopefully the others won't pick up the offender's bad habits.

Q: How do I know if an egg is too old to eat?

A: As eggs get older, they lose some moisture through the mildly porous shell. That moisture is replaced by air. A common test for the freshness of eggs is to put them in a pot filled with water. An older egg with less moisture and more air will float in the water, and the fresher ones will roll around on the bottom. Floating just means the egg is older, not necessarily that it has gone bad.

I recommend cracking the floating eggs outdoors. If they don't stink, then they are fine to eat. An egg will smell rotten if anaerobic bacteria — which survive without oxygen inside the egg — have started to decompose the egg itself. If any eggs do emit the sulfurous fumes of a rotten egg, toss them in the trash or in a rodentproof compost bin where the aerobic bacteria will quickly overcome the stinky anaerobic bacteria.

Q: Why are some eggs larger or smaller?

A: To some degree the size of the egg is determined by the size of the hen, but it is also determined by her diet and health that week. Just like people, hens can have a busy or bad week of being henpecked, and their productivity, in terms of egg size, may go down temporarily. In most cases, undersized or oddly shaped eggs are temporary issues.

Q: Sometimes there are speckles or raised spots on the eggs my hen lays. Should I worry about that?

A: Nothing to worry about. Speckles are just an indication of some variation in diet or health that week for that hen. The same goes for raised spots or oddly shaped eggs. Humans don't always look or feel their best every week: allergies, acne, and so on affect us all. But we get over it, and in most cases your hens will, too.

Q: Do hens lay eggs sitting down?

A: No. They stand up at the last minute and let the egg drop. That's why filling nest boxes with soft bedding is important.

Grounding Your Birds by Clipping Their Wings

Chickens can fly. Which can get them into trouble outside your yard. Spare everyone potential grief by clipping their wings. Feathers are like hair or nails and it doesn't hurt to clip them. You only have to do it on one wing. A clipped wing means they can't fly in a straight line, so they forgo flying. They will still be able to hop up on their roosts.

Since feathers grow back you will have to clip a wing every year. After they molt in the fall is a good time to do it, as the clipping will last for a full year.

One person can do this with some difficulty; it's much easier with two.

1. Sit on a chair and flip the chicken on its back on your lap. With its feet toward you (so you can hold the legs with one hand) and its head hanging down past your knees it will fall asleep as its tiny brain is flooded with blood.

   Alternatively, hold the bird like a football (as Chris is doing in the photo) with your forearm under its belly, your hand holding its feet, your upper arm pinning the wings to your side and its head looking out behind you.

2. Hold the leading edge of one wing and extend it like a fan.

3. With a pair of scissors clip off more than half the length of the longest 10 feathers; these are the flight feathers, which are grouped together at the front of the wing. The chicken won't feel a thing.

   Then relax knowing that your chickens won't be flying into your neighbors' yards to play lethal games with anyone's four-legged pets.

REFERENCES AND RESOURCES

Henkeepers Profiled

Hans Voerman
www.ecolodgebonaire.com

Suvir Saran
www.suvir.com

Amy Stewart
www.amystewart.com

Other Resources

WATERING SYSTEMS
The collapsible bucket holder used in the rainwater drinking fountain can be found at www.farmtek.com.

The Walden Effect
www.waldeneffect.org

Avian Aqua Miser
www.avianaquamiser.com

WINE CAP MUSHROOMS
Mushroom Mountain
www.mushroommountain.com

AUTOMATIC CHICKEN DOORS
Andrew Wells
www.youtube.com/user/andywellsinventor

Iva Biggin
http://youtu.be/mhq9mcGpeN4

METRIC CONVERSIONS

To convert	to	multiply
inches	centimeters	inches by 2.54
feet	meters	feet by 0.3048
pounds	kilograms	pounds by 0.45
teaspoons	milliliters	teaspoons by 4.93
tablespoons	milliliters	tablespoons by 14.79
cups	liters	cups by 0.24
pints	liters	pints by 0.473
quarts	liters	quarts by 0.946
gallons	liters	gallons by 3.785

To convert	to	
Fahrenheit	Celsius	subtract 32 from Fahrenheit temperature, multiply by 5, then divide by 9

METRIC EQUIVALENCES

US (inches)	Metric (centimeters)	US (inches)	Metric (centimeters)
⅛ inch	3.2 mm	11 inches	27.94 cm
¼ Inch	6.35 mm	12 inches	30.48 cm
⅜ inch	9.5 mm	13 inches	33.02 cm
½ inch	1.27 cm	14 inches	35.56 cm
⅝ inch	1.59 cm	15 inches	38.10 cm
¾ inch	1.91 cm	16 inches	40.64 cm
⅞ inch	2.22 cm	17 inches	43.18 cm
1 inch	2.54 cm	18 inches	45.72 cm
1.5 inches	3.81 cm	19 inches	48.26 cm
2 inches	5.08 cm	20 inches	50.80 cm
2.5 inches	6.35 cm	21 inches	53.34 cm
3 inches	7.62 cm	22 inches	55.88 cm
3.5 inches	8.89 cm	23 inches	58.42 cm
4 inches	10.16 cm	24 inches	60.96 cm
4.5 inches	11.43 cm	25 inches	63.50 cm
5 inches	12.70 cm	26 inches	66.04 cm
5.5 inches	13.97 cm	27 inches	68.58 cm
6 inches	15.24 cm	28 inches	71.12 cm
6.5 inches	16.51 cm	29 inches	73.66 cm
7 inches	17.78 cm	30 inches	76.20 cm
7.5 inches	19.05 cm	31 inches	78.74 cm
8 inches	20.32 cm	32 inches	81.28 cm
8.5 inches	21.59 cm	33 inches	83.82 cm
9 inches	22.86 cm	34 inches	86.36 cm
9.5 inches	24.13 cm	35 inches	88.90 cm
10 inches	25.40 cm	36 inches	91.44 cm

ACKNOWLEDGMENTS

Thanks to my agent, Sally McMillan, for good conversations, good feedback, and perseverance in the face of adversity.

This well-crafted book wouldn't be in your hands without the talents, collegiality, and hard work of Deb Burns, Corey Cusson, Deborah Balmuth, Michaela Jebb, Liz Nemeth, Alex Tricoli, Michael Gellatly, Janet Renard, Jim Sollers, Karen Martin, Alee Moncy, Melinda Slaving, Ilona Sherratt, Megan Posco, and Sarah Armour.

Thanks to henkeepers and helpers Katie Ford, Johnny Ford, Bob Davis, Judy Davis, Beverly Norwood, Woody Collins, Michelle Old, Hillary Nichols, Micaela Stanton, Casey Stanton, Felipe Witchger, Michele Kloda, Cyrus Dastur, Alena Dastur, Amy Stewart, Suvir Saran, Hans Voerman, Tradd and Olga Cotter, as well as the many henkeepers who've been on the Bull City Coop Tour, the Greenboro Coop Loop, and the Raleigh Tour de Coop, for sharing their coops, chickens, and stories with me.

A big nonverbal thank-you to my coworkers on Monkey Island, Turtle Island, and New Light Farm for showing me so much about the minds, bodies, and souls of our fellow animals.

A plumb, level, and square thank-you to carpenters Mark Marcoplos, Sam Dennis, Dave Richardson, Bill Wallace, Jamie Wallace, John Carroll, and so many others, for teaching me how to build anything, as well as how to finish a project with the same number of fingers with which I began it.

A four-dimensional thank-you to my design professors Tracy Traer, Curtis Brooks, Fernando Magallanes, Denis Wood, and Will Hooker for breaking my brain open so that anything I could imagine I could also design and build.

A six-times-a-year thank-you to Roger Sipe, my editor at *Chickens* magazine, for taking a chance on a type of column that had never appeared in a poultry magazine before. Many of the photos and earlier drafts of some of the text appeared in my "Coop Builder" column in that magazine.

And special thanks to my first editor back in the early 1990s, Gillian Floren at North Carolina's *Indy Week*, for Rolfing some proper posture into my then-flabby, flimsy, freshman writing style.

ABOUT THE AUTHOR

As a highly successful college dropout in the 1980s, Frank Hyman worked with loggerhead sea turtles, beef cattle, and free-ranging rhesus monkeys. He also lasted one day at a factory egg farm. He eventually earned a B.S. in horticulture and design from North Carolina State University on the "Jethro Bodine" plan. Meaning he invested 8 years earning his 4-year degree.

Frank's need for an outdoor life led him to become a carpenter, stonemason, sculptor, and metalworker with 40 years of experience in home, farm, and garden construction on two continents. In addition to many arbors, stone walls, treehouses, and other structures, he's built a doghouse with a green roof, a cat feeding station with a copper roof, a chicken coop with a pagoda roof, and a tractor shed in Provence made from tree trunks, bent nails, and other salvaged materials.

He writes columns for four magazines: *Chickens, Paleo, Hobby Farm,* and *Horticulture*. His writing has also appeared in the *New York Times,* the *Wall Street Journal, Modern Farmer, Organic Gardening, Fine Gardening, American Gardener, Carolina Gardener, Bicycle Times,* and *Backyard Poultry.*

He and his wife, Chris, helped lead a group called Healthy Eggs in Neighborhoods Soon (HENS) in their hometown's successful campaign to legalize backyard hens in 2009. They won a unanimous victory on the Durham City Council, and they have Healthy Eggs in Neighborhoods Now.

Frank leads DIY coop design/build classes twice a year based on his Hentopia coop and hen habitat system. Learn more at www.hentopiacoops.com.

Frank lives and gardens with his wife, Chris Crochetiere; their yellow Labrador, Abbey; a school of goldfish; and a small flock of hens at Bayleaf Cottage in Durham, North Carolina.

INDEX

Page numbers in *italic* indicate photos or illustrations.

Spread Your Wings with More Books from Storey

BY JENNA WOGINRICH, PHOTOGRAPHY BY MARS VILAUBI

Absolute beginners will delight in this photographic guide chronicling the journey of three chickens from newly hatched to fully grown, highlighting all the must-know details about chicken behavior, feeding, housing, hygiene, and health care.

BY JUDY PANGMAN

This collection of 45 hen hideaways promises to spark your imagination and inspire you to begin building. Everything you need is detailed in the basic plans, elevation drawings, and building ideas for these original coops.

BY MELISSA CAUGHEY

Learn how your birds communicate, understand the world around them, and establish roles within the flock. Personal observations and accessible scientific findings help answer questions like *Do chickens have names for each other?* and *How do their eyes work?*

BY KEVIN McELROY & MATTHEW WOLPE

Complete plans for 14 contemporary chicken coops marry function and style. Whether you're an expert carpenter or new to building, step-by-step instructions and detailed illustrations will ensure that your project is a success.

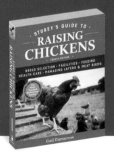

BY GAIL DAMEROW

Here's all the information you need to raise happy, healthy chickens. Learn about everything from selecting breeds and building coops to incubating eggs, hatching chicks, keeping your birds healthy and safe from predators, and much more.

31901064109814

Join the conversation. Share your experience with this book, learn more about Storey Publishing's authors, and read original essays and book excerpts at storey.com. Look for our books wherever quality books are sold or call 800-441-5700.